Just War Theory and Artificial Intelligence

Two of the most crucial issues facing humanity in the contemporary world are artificial intelligence and armed conflict. This book tackles them both, bringing them into conversation with one another and creating an understanding of their relationship. New developments in technology, especially within weapons systems, have challenged our understandings of just war theory, thus changing the nature of the questions to be asked of and by those engaged in conflict. Covering topics such as drone warfare and the use of wargaming by combatants, this collection seeks to answer those questions with essays from leading experts in the field. It will be a valuable resource for those interested in both the development of artificial intelligence generally and warfare specifically.

Maria Power is a senior research fellow at the Las Casas Institute for Social Justice, Blackfriars Hall, University of Oxford. Her work focuses mainly on peacebuilding within divided communities in Northern Ireland. Her most recent publications include *Catholic Social Teaching and Theologies of Peace in Northern Ireland: Cardinal Cahal Daly and the Pursuit of the Peaceable Kingdom*, (Routledge, 2021) and *Violence and Peace in Sacred Texts*, (Palgrave, 2023). She is currently writing a monograph exploring the role of the Bible in the conflict in Northern Ireland which will be published by Routledge in 2027.

Maggi Savin-Baden is a professor and senior research fellow at Las Casas Institute for Social Justice Blackfriars Hall, University of Oxford. She has authored, coauthored and edited 29 books in the areas of innovative learning, digital fluency, artificial intelligence, the postdigital, digital afterlife, pedagogical agents, qualitative research methods, problem-based learning and the metaverse. She currently coeditor of the Metaverse book series. In her spare time, she runs, bakes, climbs, does triathlons and wild swimming.

Chapman & Hall/CRC
Artificial Intelligence and Robotics Series
Series Editor: Roman Yampolskiy

An Introduction to Universal Artificial Intelligenc
Marcus Hutter, Elliot Catt, and David Quarel

Transcending Imagination: Artificial Intelligence and the Future of Creativity
Alexander Manu

Responsible Use of AI in Military Systems
Jan Maarten Schraagen

AI iQ for a Human-Focused Future: Strategy, Talent, and Culture
Seth Dobrin

Federated Learning: Unlocking the Power of Collaborative Intelligence
Edited by M. Irfan Uddin and Wali Khan Mashwan

Designing Interactions with Robots: Methods and Perspectives
Edited by Maria Luce Lupetti, Cristina Zaga, Nazli Cila, Selma Šabanović, and Malte F. Jung

The Naked Android: Synthetic Socialness and the Human Gaze
Julie Carpenter

AI: Unpredictable, Unexplainable, Uncontrollable
Roman V. Yampolskiy

De Novo Quantum Cosmology with Artificial Intelligence: Applications of Formal Autoencoders
Ariel Fernández

Just War Theory and Artificial Intelligence: Challenges and Consequences
Edited by Maria Power and Maggi Savin-Baden

For more information about this series please visit: ***https://www.routledge.com/Chapman--HallCRC-Artificial-Intelligence-and-Robotics-Series/book-series/ARTILRO***

Just War Theory and Artificial Intelligence

Challenges and Consequences

Edited by
Maria Power
Maggi Savin-Baden

CRC Press
Taylor & Francis Group
Boca Raton London New York

CRC Press is an imprint of the
Taylor & Francis Group, an **informa** business
A CHAPMAN & HALL BOOK

Designed cover image: Shutterstock_2532747835

First edition published 2026
by CRC Press
2385 NW Executive Center Drive, Suite 320, Boca Raton FL 33431

and by CRC Press
4 Park Square, Milton Park, Abingdon, Oxon, OX14 4RN

CRC Press is an imprint of Taylor & Francis Group, LLC

ISBN: 978-1-032-57622-0 (hbk)
ISBN: 978-1-032-57624-4 (pbk)
ISBN: 978-1-003-44025-3 (ebk)

DOI: 10.1201/9781003440253

Typeset in Times
by SPi Technologies India Pvt Ltd (Straive)

Contents

Contributors

Christian Nikolaus Braun is a lecturer in the Defence Studies Department at King's College London. He was formerly a Radboud Excellence Initiative Fellow at Radboud University, and a Senior Lecturer in the Defence and International Affairs Department at the Royal Military Academy Sandhurst. His primary area of research is the ethics of war and peace. It is his ambition to bring to bear the wisdom of the just war tradition on ethical issues that confront us today. For example, in his monograph *Limited Force and the Fight for the Just War Tradition* (Georgetown University Press, 2023), Christian employs a neoclassical reading of just war that is grounded in the thought of Thomas Aquinas to argue about so-called uses of force-short-of-war. His research has also been published in numerous peer-reviewed journals, including *Ethics & International Affairs*, *Global Studies Quarterly*, *International Affairs*, *International Relations*, *International Theory*, the *Journal of International Political Theory*, and the *Journal of Military Ethics*.

Malcolm Brown is Director of Faith and Public Life for the Church of England and a policy adviser to the Archbishop of Canterbury. He has been a parish priest, Director of a think tank on theology and economics and has taught Ethics and Practical Theology for a number of universities. He teaches ethics for students at the University of Bath's Centre for Doctoral Training in Accountable, Responsible and Transparent AI (ART-AI). He is the author of *Tensions in Christian Ethics* (2010) and *Anglican Social Theology* (2014) among other books and articles etc.

David J. H. Burden has been a wargamer and wargame designer for around 50 years – although ironically not for the 10 years he spent in the British Army. For the last 20 years David has run Daden Limited, helping organisations explore and exploit the social and commercial potential of using conversational AI and virtual worlds, delivering over 100 projects for clients across the globe, including over two dozen projects for UK MOD. David started his part-time PhD on wargaming urban conflict in early 2022, under Dr John Curry at Bath Spa University. He is currently researching and designing games on the urban conflict, including The Urban Calculus which looks specifically at the linkages between fighting, casualties, reverberating damage, global opinion, societal support and government will – all key aspects of a Just War. David is an ex-Royal Signals officer, a Chartered European Engineer and is also series co-editor for Taylor & Francis on their Metaverse Series of books, and co-author of both *The Metaverse: A Critical Introduction*, and *The Military Metaverse*.

Jovana Davidovic is an Associate Professor of Philosophy at the University of Iowa, where she also holds a secondary appointment at the Center for Human Rights and the Law School. Professor Davidovic is also a Senior Research Fellow at the United States Naval Academy and a Chief Ethics Officer for BABL AI, an AI and algorithm research consultancy focusing on AI risk assessments and AI auditing. Professor Davidovic's research focuses on ethics of AI and algorithms, with a focus on ethics of AI-enabled weapons systems and AI risk assessments. Her previous work includes papers on topics in military ethics, international law, human rights and AI ethics.

James Garcia is a theology PhD student at Villanova University. His writings include a journal article titled Fighting for Peace: Church Thinking on War from Vatican II to the war on Ukraine (Journal for Peace and Justice Studies, 32 (1/2)). His research interests include the impact of drones, precision-guided munitions, and AI-enabled weapon systems on twenty-first century military ethics and the law of armed conflict. He is also exploring whether jus ad bellum concepts may be useful guides for not only justifying a war but ending it.

Matthew J. Gaudet is an associate teaching professor and director of Ethics Programs for the School of Engineering at Santa Clara University. His research lies at the intersection of moral theology and social science, with a particular interest in the ethical questions of technology, war and peace, disability, and the university. He is a lead author of *Encountering Artificial Intelligence: Ethical and Anthropological Investigations* (Wipf & Stock, 2024), co-author of *Eight Theories of Justice: Perspectives from Philosophical and Theological Ethics* (Fortress, 2025), and a co-editor of three issues of the *Journal of Moral Theology*, including a 2022 special issue on Artificial Intelligence.

Noreen Herzfeld is the Nicholas and Bernice Reuter Professor of Science and Religion at St. John's University and the College of St. Benedict and a Senior Research Associate at the Institute for Philosophical and Religious Studies, in Koper, Slovenia. She holds degrees in Computer Science and Mathematics from The Pennsylvania State University and a Ph.D. in Theology from The Graduate Theological Union, Berkeley. Recent works she has written or co-edited include *The Artifice of Intelligence: Divine and Human Relationship in a Robotic Age*; *In Our Image: Artificial Intelligence and the Human Spirit; Technology and Religion: Remaining Human in a Co-Created World*; *Encountering AI: Ethical and Anthropoligical Investigations; Religious and Cultural Implications of Technology-Mediated Relationships in a Post-Pandemic World*; and *Religion and the New Technologies*.

Bernhard Koch studied philosophy, logic and scientific theory in Munich and Vienna. In 2006 he received his doctorate in ancient philosophy on Cicero. Between 2007 and 2024 he was project manager and deputy director of the Institute for Theology and Peace in Hamburg. He is an adjunct professor of theological ethics at the University of Freiburg and has held various teaching positions in Frankfurt and Hamburg. In 2025 he took up a new position as a researcher at the Institute for Military Medical Ethics of the Bundeswehr in Munich. In addition to general ethics, his research focuses on the ethics of ius in bello, technology ethics and the ethics of AI.

Robert H. Latiff is an independent contractor and consultant, providing technology advice to corporations, universities, and the military. He is an adjunct faculty member at the University of Notre Dame and was recently a Distinguished Fellow In-Residence at Notre Dame's Institute for Advanced Study. Dr. Latiff retired from the U.S. Air Force as a Major General in 2006. His last active-duty military assignment was at the US National Reconnaissance Office. In a prior assignment he served as the commander of the US Space Command Cheyenne Mountain Operations Center. After retirement from the USAF, Dr. Latiff spent three years as Chief Technology Officer in Science Applications International Corporation's (SAIC) Geospatial Intelligence business. General Latiff received his commission from the Army ROTC program at the University of Notre Dame and served in Europe as an infantry platoon leader and then commander of an Army tactical nuclear weapons unit. Prior to transferring to the USAF, he was as Assistant Professor of Engineering at the US Military Academy. He received his Ph.D. in materials science and B.S. in physics from the University of Notre Dame and is a graduate of the National Security Fellows Program at Harvard's JFK School of Government. Dr. Latiff is the author of *Future War: Preparing for the New Global Battlefield* (Knopf, 2017) and *Future Peace: Technology, Aggression, and the Rush to War* (Notre Dame Press, 2022).

James McEvoy is a serving officer in the Royal Corps of Signals. He has hands-on experience with deploying AI for the Army and incorporating it into established military frameworks, coupled with a strong interest in the ethics of AI. This has included work on AI being used as deterrence, its impact on leadership theory, and the ethics of using AI in military situations. He has written several articles on the topic and delivered talks to both military, industry, and academic colleagues, as well as working with the UKRI Centre for Doctoral Training in Accountable, Responsible and Transparent AI (ART-AI).

Eamonn O'Neill is Head of Computer Science at the University of Bath and Director of the UKRI Centre for Doctoral Training in Accountable, Responsible and Transparent AI (ART-AI). He has led interdisciplinary research projects and technology transfer activities over many years and has published numerous papers in leading academic conferences and journals. His research covers the range of technological, cognitive and social challenges and opportunities that interactive systems offer. Key topics include innovative forms of human–technology interaction and technology-mediated human–human interaction, including interaction with 'intelligent' machines, software and systems.

Maria Power is a Senior Research Fellow at the Las Casas Institute for Social Justice, Blackfriars Hall, University of Oxford. Her work focuses mainly on peacebuilding within divided communities in Northern Ireland. Her most recent publications include *Catholic Social Teaching and Theologies of Peace in Northern Ireland: Cardinal Cahal Daly and the Pursuit of the Peaceable Kingdom*, (Routledge, 2021) and *Violence and Peace in Sacred Texts*, (Palgrave, 2023). She is currently writing a monograph exploring the role of the Bible in the conflict in Northern Ireland which will be published by Routledge in 2025.

Mitt Regan is McDevitt Professor of Jurisprudence and Coordinator of the Program on Law, Ethics, and International Security at Georgetown Law Center, and is a Senior Fellow at the Stockdale Center for Ethical Leadership at the U.S. Naval Academy. He also is a member of an expert group advising the Defense Advanced Research Projects Agency (DARPA) on legal and ethical dimensions of the military use of AI. He works in the fields of the law and ethics of military applications of artificial intelligence, international law on the use of force, international human rights law, international criminal law, and military ethics. He is the author most recently of *Drone Strike: Analyzing the Impacts of Targeted Killing* (Palgrave Macmillan 2022), and co-editor of *Hybrid Threats and Grey Zone Conflict: The Challenge to Liberal*

Democracies (Oxford University Press 2024); *Between Crime and War: Hybrid Legal Frameworks for Asymmetric Conflict* (Oxford University Press 2023); and *National Security Intelligence and Ethics* (Routledge 2022).

Maggi Savin-Baden is a Professor and Senior research fellow at Las Casas Institute for Social Justice Blackfriars Hall, University of Oxford. She has authored, coauthored and edited 29 books in the areas of innovative learning, digital fluency, artificial intelligence, the postdigital, digital afterlife, pedagogical agents, qualitative research methods, problem-based learning and the metaverse. She currently co-editor of the Metaverse book series. In her spare time, she runs, bakes, climbs, does triathlons and wild swimming.

Zak Savin-Baden completed his degree in Computer Science from Birmingham City University gaining a first. He has since completed his Master's Degree in Artificial Intelligence and currently working towards a PhD. In his spare time, he is a gamer, programmer and enjoys making 3D Models with his 3D Printer.

Kenneth Wilkinson-Roberts (he/they) is currently a final year PhD student at Lancaster University's Politics, Philosophy and Religion Department. With a background in Religious & Peace Studies, he researches the intersections between ministry, technology and the future of War. Currently, Kenneth's current research explores the ways that emerging technologies affect the pastoral lives of military personnel.

Introduction

Maria Power and Maggi Savin-Baden

The nature of warfare is constantly evolving, with each new innovation bringing fresh ethical challenges that change our understandings of war. If the 12th century was the century of the crossbow, the 13th that of the cannon, and the 20th that of nuclear arms, then the 21st century is certainly the era of artificial intelligence (AI) in warfare, especially as 'military applications of AI are receiving some of the greatest investments in the field' (AI Research Group of the Centre for Digital Culture, 2023, 199). Just as these new technologies have initiated transformations in the nature of war, they have also 'led to a re-evaluation not just of how the principles of Just War are to be applied, but also how they are interpreted' (Blanchard and Taddeo, 2024, 708). Indeed, emerging military technologies in the field of AI are considered so problematic that Pope Francis called for them to be banned (Watkins, 2024); and in 2024 Archbishop Ettore Balestrero stated that 'it is profoundly distressing that, adding to the suffering caused by armed conflicts, the battlefields are also becoming testing grounds for more and more sophisticated weapons' (Watkins, 2024) Additionally, the Holy See has called for a deliberate use of language in relation to the use of AI on the battlefield that employs the concepts of human dignity and a consideration of values and duties. There has been an academic focus on the relationship between just war teaching and artificial intelligence for about twenty years. Critiques have come from both a secular and religious perspective. As Brian Stilner puts it 'just war principles can address drones and robots along the same lines as crossbows, chemical weapons, and nuclear weapons, because behind the use of all of them are very similar motivations.' (Stiltner, 2015, 23) This collection of essays seeks to a survey of the current debate surrounding just war and AI.

The essays have been presented in thematic groups. The first group explores artificial intelligence in relation to the conditions of war set down in just war teaching. In 'Jus Ante Bellum and AI-enabled Weapons', Jovana Davidovic and Mitt Regan discuss the use of AI in relation to states' preparations for war. They argue that jus ante bellum requires that states must meet two obligations: first, that AI weapons are safe, reliable and lawful; and second, that the development of such weapons does not trigger a sense of insecurity on the part of other states. Robert Latiff's 'Artificial Intelligence in decisions on Armed Conflict' maintains that issues relating to command control and decision making when using AI weaponry falls under the remit of jus ad bellum. He argues that ceding to technology in key decisions in war violates just war teaching. In 'Jus in Bello: AI is not just another weapon', Malcolm Brown, James McEvoy, and Eamonn O'Neill, raise the question of how just war ethics can be incorporated into military technologies if the ethical assumptions of their designers and developers cannot be assumed. The second set of essays evaluates the impact of specific technological developments on just war theory. Taking Pope Francis's technocratic paradigm as his starting point, Matthew J Gaudet examines how lethal autonomous weapons systems impact upon Catholic just war teaching from the perspective of jus in bello criteria. James Garcia's 'Above and Beyond: Artificial Intelligence and Just War Theory' argues that developments in AI, and in particular the use of drones in warfare, strain the limits of just war thinking. 'On the morality of fully-autonomous armed drones' by Christian Nikolaus Braun and Bernhard Koch focuses on the use of drones from the perspective of human dignity, a key teaching of the Roman Catholic church. Noreen Herzfeld's 'The paradox of distance: AI, violence and embodiment on and off the battlefield' asks whether new technologies of lethal automated weapons systems and social media promote violence or nonviolence and whether they strengthen human solidarity and dignity or fracture them. The third set of essays looks at specific case studies. Kenneth Wilkinson-Roberts explores the consequences of AI for military chaplaincy in 'Actions and Accidents in the Future Battlefield'. Whilst in 'Wargaming and the Just War', David Burden examines the extent to which wargaming can be used to support the planning and analysis of warfare through the three key stages of just war teaching. Finally, Zak Savin-Baden's chapter, 'Fighting just war in a digital realm' explores the ways in which just war teaching is integrated into games such as minecraft and shows how just wars can be fought in the digital realm.

The aim of this collection is to present the main themes that are being debated and considered by scholars of just war teaching and artificial intelligence. As artificial intelligence becomes more ubiquitous both in warfare and in society in general, it is hoped that this collection of essays will provide a starting point for a consideration of one of the key concerns for those considering the ethics of war and peace in the 21st century.

REFERENCES

AI Research Group of the Centre for Digital Culture. (2023). *Encountering Artificial Intelligence: Ethical and Anthropological Investigations*. Theological Investigations of Artificial Intelligence Book Series, Journal of Moral Theology. https://jmt.scholasticahq.com/article/91230-encountering-artificial-intelligence-ethical-and-anthropological-investigations

Blanchard, A., & Taddeo, M., (2024). Autonomous Weapons Systems and *jus ad bellum*. *AI & Society*, 39, 705–711.

Stiltner, B., (2015). 'A Taste of Armageddon: When Warring is done by Drones and Robots'.

Watkins, D., (2024). 'Holy See: Technology should better human life, not take it.' *Vatican News*. https://www.vaticannews.va/en/vatican-city/news/2024-08/holy-see-un-killer-robots-technology-human-life.html

Winright, T., & Johnston, L., (n.d.) *Can War be Just in the 21st Century? Ethicists Engage the Tradition*. New York: Orbis. 14–28.

1 Jus Ante Bellum and AI-enabled Weapons

Jovana Davidovic and Mitt Regan

INTRODUCTION

The prospect of weapons whose operations incorporate artificial intelligence (AI) raises important issues relating to traditional just war concerns about when States in the international community may justifiably resort to war and how they may use force once war has begun. It is important to appreciate, however, that whether a State goes to war and how it conducts it may be shaped by how it prepares for the possibility of war. Certain forms of military preparedness may make war less or more likely or may enable or foreclose ways in which a State fights war. recognition of this has led to an emerging strand of just war thought known as *jus ante bellum*. It asks: what ethical principles should guide how a State prepares for war?

Van der Linden offers a useful approach to this inquiry (2010). As he puts it, *jus ante bellum* asks 'whether the military preparation of the country is such that it is conducive to the country resorting to force only when justice is on its side, as well as to executing and concluding war in the just manner. How should we prepare for the possibility of military conflicts so that wars will be only justly initiated, executed, and concluded?' (van der Linden, 2010, pp. 1–2). In other words, just preparation for war minimizes the risk that a State will violate the principles of *jus ad bellum* and *jus in bello*.

This chapter examines what *jus ante bellum* requires of States with respect to the development and fielding of AI-enabled weapons (AEW). We use the term 'development' to refer to the process of ensuring that a weapon meets specifications and requirements necessary to make it available for use. We use the term 'fielding' to refer to making it available to armed forces for their use. Because of the distinctive risks these weapons can pose, jus ante bellum requires that a State: (1) conduct a rigorous testing, evaluation, verification, and validation (TEVV) and legal review process in developing an AEW to ensure that it can be used safely, reliably, and lawfully before it is fielded; and (2) that it engage in development and fielding of these weapons in ways that avoid triggering a security dilemma in which other States feel the need to field AEW without engaging in this process.

The section that follows describes the key risks of AEW. The section after that discusses the critical role of rigorous TEVV in reducing these risks. The chapter then discusses how even a State that conducts rigorous TEVV may develop and field AEW in ways that lead other States to hasten fielding AEW without sufficient TEVV. We

DOI: 10.1201/9781003440253-1

suggest how States might avoid this risk. A concluding section summarizes the importance of using the *jus ad bellum* framework to determine what ethics requires of States that develop and field AEW as one means to protect themselves in case of war.

To clarify the focus of this chapter, we note that some argue that AEW can never be used ethically and therefore should be prohibited. Others maintain that only AEW with certain features should be permitted. Either camp may argue that *jus ante bellum* supports its position. We bracket and take no position in those debates for purposes of this chapter. Our argument is that, assuming that States continue to develop AEW, regardless of what features they have, *jus ante bellum* requires that they conduct rigorous TEVV and take steps to avoid triggering a security dilemma.

RISKS OF AI-ENABLED WEAPONS

The development and use of AEW may pose the risk of violations of *jus ad bellum* and *jus in bello* for several reasons. First, *testing* systems for reliability and safety is a major challenge because it is impossible fully to replicate before fielding the unpredictable conditions of warfare. Second, weapon systems can be vulnerable to *adversarial attacks* that distort the information that determines how they operate. Warfare by definition involves conflict between adversaries, so this risk seems especially high. Third, systems at this point tend to be *brittle*, in the sense that they are not able to function effectively outside the specific set of circumstances for which they are trained. Fourth, it is not always clear with AEW what counts as being 'outside' the circumstances for which they are trained. Fifth, small, sometimes imperceptible, changes to the environment and context of use can significantly affect performance. This makes it challenging to completely predict behaviour that could result in significant harm to innocent persons or one's own force. Sixth, it can be challenging to determine when a system has ventured into this situation and therefore when it is necessary to disable it. Finally, a system may not be able to provide an *explanation* of its analysis and recommendations in terms that are comprehensible to a human operator. This *opacity* can make it difficult for humans to exercise effective judgment, thus creating the risk of deferring to machine outputs without full situational awareness.

With respect to *jus ad bellum*, these features of AEW could result in States fielding and using systems that are less flexible than conventional weapons, which lack sensitive contextual awareness of likely human intentions. This limitation 'may particularly be challenging in pre-conflict crisis situations, where tensions among nations run high,' and contextual human judgment can be crucial in lessening the risk of escalation (Horowitz & Scharre, 2021, p. 8).

If a system acts in a destructive way, other States may not be certain if this is intentional. This may lead to escalation of conflict if those States assume that they must ascribe hostile intention in order to protect themselves. Finally, systems' 'intrinsic vulnerability to unexpected interactions or operational accidents raises the spectre of inadvertent escalation into a 'flash war' between autonomous military systems, similar to the algorithmic flash crashes already observed in the financial sector' (Scharre, 2018). All these scenarios pose the risk that States will resort to war without just cause or may not meet other requirements of the *jus ad bellum*.

With respect to *in bello* violations, 'minor tactical missteps or accidents that are part and parcel of military operations in the chaos and fog of war, including fratricide, civilian casualties, and poor military judgment, could spiral out of control and reach catastrophic proportions before humans have time to intervene' (Horowitz & Scharre, 2021, p. 5). If so, processes for deliberation based on human judgment would be unavailable, which could mean intensification of warfare resulting in greater suffering and larger loss of innocent life. This risk would be exacerbated by the interaction between competing AEW systems that could result a cycle of attacks and counterattacks at a speed that humans could not control. These risks underscore the crucial importance of rigorous review of any AEW on which a State may rely in preparing for the possibility of war. The next section discusses what is involved in such review.

TESTING, EVALUATION, VERIFICATION AND VALIDATION

As we have described, fielding AEW that have not been rigorously tested for safety and reliability increases the risk of unjust resort to war and of harm to innocent persons during war. To avoid this, fielding should be preceded by development that includes a rigorous process known as testing, evaluation, verification, and validation (TEVV). TEVV is a process that includes testing- assessing performance, reliability and safety of an engineered system or a product; evaluation, a systematic determination of the extent to which the technology has met some specified criteria; verification, provision of evidence that a system performs as intended and in accord with the performance criteria; and validation, confirmation from objective evidence that requirements have been met for a specific intended. This process, drawn from systems engineering, is designed to assess the future performance of new technology and the risks that it may pose. The TEVV process seeks to provide assurance that technology will work as expected, which generates what Roff and Danks call trust based on predictability. (Roff & Danks, 2018, p. 3).

Because a weapon can cause significant harm, TEVV of weapons also aims to provide what Roff and Danks call trust based on values: confidence that a weapon will operate in a way that is consistent with certain ethical demands, such as the legal requirement of not targeting civilians. When we refer to TEVV in this chapter, we include review of weapons to ensure that they can be used lawfully.

The paradigm of values-based trust is interpersonal relationships, in which trust reflects confidence that another person will act ethically in unpredictable future situations because we know the 'values, principles, beliefs and motives that guide the trustee's choices and actions' (Roff & Danks, 2018, 7). Such understanding is based on repeated interactions with another in which we 'perform our own risk calculations about what to internalize about the other agent's mindset and how vulnerable to make ourselves' (Roff & Danks, 2018, 10).

In human-machine teams, however, the 'values' of the machine may be inscrutable to humans. Roff and Danks argue that AEW pose a fundamental tension: the extent that an AEW 'actually *learns* and adapts to its environment in operationally ways will be *inversely* proportional to the extent to which the human team members can identify and internalize its values and preferences' (Roff & Danks 2018, p. 11).

The more advanced an AEW system, the greater the need to have not only the reliability-based trust that we have in tools, but the kind of values-based trust that we have in other humans.

A weapon system TEVV thus must seek to generate the right kind of calibrated trust of both types in commanders who decide to deploy a weapon and operators who use it (Flournoy et al., 2020). Trust is *calibrated* when 'the warfighters' operational reliance aligns with the system performance for the context' (Pinelis, 2022). Trust is the *right kind* when it is grounded both in predictability and conformity with values (Roff & Danks, 2018). As the discussion below describes, there are features of AEW that can make it difficult for TEVV to generate this trust in such weapons.

Challenges

AEW present distinctive challenges for TEVV. We discuss these challenges below and suggest how TEVV should respond to them in order to satisfy the *jus ante bellum* requirement to provide assurances of safety, operational effectiveness, and conformity with ethical values.

Appropriate Unit of Analysis

It is harder to define the appropriate unit of analysis for TEVV for an AEW than for a conventional weapon. First, the same algorithm can be utilized across a range of weapon system applications. An object recognition algorithm, for instance, might be used to enable autonomous navigation in a tank, or to inform targeting by distinguishing between objects that are weapons and those that are not.

Second, algorithms often function within a system of systems. This means that several different algorithms may each provide inputs for one another. Thus, for instance, an object recognition algorithm could provide input to a decision support algorithm that collates and processes inputs from other algorithms and presents alternative courses of action to a commander.

Third, the ability of an algorithm to function properly greatly depends on the interaction between humans and machines. This means that it may be necessary to test AEW with humans who have different types of training in order to gain assurance of the weapon's performance (Flournoy et al. 2020).

All this means that it is not always easy to identify and circumscribe the appropriate unit of analysis for a test or a risk assessment. For instance, is the test for an algorithm in a particular application? If so, how much of the testing must be repeated for other applications?

Fitness for Purpose

AI solutions and AEW are built for a purpose. A pre-requisite for good TEVV is clearly stating that purpose and being able to assess whether the tool serves it as well as potential alternatives, and whether 'ground truth' benchmarks for evaluating it accurately reflect contexts in which an AEW weapon will be used. The consequences of using the system in the wrong context are significant. and what counts as a 'wrong' context is much more difficult to discern for an AEW than for other types of systems.

Generalization from Testing

Generalizing and extrapolating from test results is challenging for many AEW because of the difficulty in anticipating all the conditions under which these weapons will operate. It is true that conventional weapons present a similar obstacle to some extent – we can test only a fraction of the settings in which a weapon may operate. AEW, however, perform extremely complex tasks, they do so in radically unpredictable environments, and they provide 'non-deterministic, dynamic responses to those environments.' (Wojton et al., 2021, p. 4). All of this makes the range of potential test scenarios immense. This means that compared with conventional weapons, we can generalize with much less confidence about performance across varied environments (Pinelis, 2022).

Unpredictable Failures

The fact that AEW are especially complex and often opaque means that their failures are harder to predict and more difficult to understand than those of conventional weapons. This complicates the ability of TEVV to identify situations in which the weapon should operate. Systems that rely on deep learning and those that are components of larger systems are particularly likely to present these challenges. This leads some observers to suggest that 'explainability' may be a prerequisite for adequate TEVV, and thus for fielding a weapon (Flournoy et al., 2020, p. 20).

Piecemeal Approach to Development

Unlike conventional weapons, AEW tend not to be custom-built for the military. They (and specifically their AI components) more likely are assembled piecemeal from various sources. This is because cutting-edge AI development generally is occurring in the private sector, and because AI often is utilized to solve specific narrow problems. An AEW thus will come from a variety of sources. This means that the TEVV process for a given AI component will not provide assurance of its performance for military use. While the U.S. Department of Defense, for example, may be able to use some commercial TEVV tools, in many cases 'industry methods will not be applicable, given the safety-critical application and unique classification of DOD data' (Flournoy et al., 2020, p. 17).

Dynamic and Open AI Systems

Finally, some AEW may eventually be online learning systems – that is, models that take in data from the environment and update the algorithmic basis of the model as it is being used. This means that at some point the system is effectively a new tool and elements of testing need to be repeated. This might, for example, require 'run-time' monitoring' while a system is operating, in order to intervene when it is about to engage in potentially harmful behaviour, or 'is making a bad decision or misinterpreting its environment' (Haugh et al., 2018, p. 3–6).

ADAPTING TEVV FOR AEW

This section discusses ways in which TEVV may be revised, given the above challenges, to ensure that AEW can meet the requirements of *jus ante bellum*.

ONGOING AND INTEGRATED TEVV OVER THE LIFE OF A SYSTEM

TEVV for an AEW should occur throughout the life cycle of the weapon. This approach will enhance the ability to anticipate problems as the weapon encounters new circumstances, and will improve the ability to set limits on the operational environments in which it may be used. This will increase the likelihood that a weapon is developed and used in ways that meet the requirements of *jus ante bellum*.

PRINCIPLED AND MODULAR APPROACH TO TEVV

There should be clear rules to determine which changes to application, use, context, or training data require or trigger a new TEVV for one of more components of a system, or for the system as a whole. We might call this 'modular' TEVV: not every change to any part of an AI system requires a completely new TEVV. Clear rules should indicate what aspects of TEVV need to be repeated under which circumstances. This is because a weapon's performance depends heavily on the operational environment, and new operational environments, significant changes in an environment, or updates to a weapon, may require new TEVV. Robust TEVV needs not only to assess performance in appropriate operational environments, but also to define what environments are appropriate for use of a weapon. This is much harder for AEW than for conventional weapons. It may be, for example, that an object recognition model works very well in one climate or geographic region, but not in another. Thus, both new training data and potentially new operational environments should trigger the requirement that some or all elements of the TEVV process be conducted anew.

INTEGRATION OF TEVV AND LEGAL WEAPONS REVIEW

Article 36 of Additional Protocol I to the Geneva Conventions requires review of a weapon to determine whether its use 'in some or all circumstances' would violate international law. Testing and technical assessment traditionally have been conducted prior to and provide evidence for legal weapons review. Because of the challenges we describe, Vestner and Rossi suggest that it may be necessary to incorporate legal requirements into the technical specifications of AEW (Vestner & Rossi 2021). This means that legal review would need to occur in tandem with the TEVV process. Under this approach, operational circumstances that trigger the need for additional TEVV may also trigger the need for additional legal weapons review.

Better Testing Data and Gradual Fielding

Many algorithms relevant to weapons systems, such as object recognition or decision augmentation algorithms, are trained and validated in simulated environments. Simulation-based testing data, however, will often be inadequate when the risks of fielding a weapon are especially high. In these cases, data sets based on actual conditions are preferable because they can increase commanders' and operators' ability to trust a system in high-risk operational environments.

It thus will be crucial in many cases that an AEW be tested in actual rather than simulated environments, and that such systems be fielded only gradually. '[A] strategy of graded autonomy (slowly stepping up the permitted risks of unsupervised tasks, as with medical residents) and limited capability fielding (only initially certifying and enabling a subset of existing capabilities for fielding) could allow the services to get at least some useful functionality into warfighters' hands while continuing the T&E process for features with a higher evidentiary burden' (Wojton et al., 2021, p. 20).

Differing Purposes of Transparency and Explainability

The transparency and explainability required to conduct the TEVV process may differ from the transparency and explainability necessary for its operation. TEVV, for instance, requires transparency and explainability that enables, among other things, the definition of the operating envelope, that is, the set of conditions under which we expect the system to perform in expected ways. An operator, by contrast, may require transparency and explainability that only enables an informed judgment about the appropriate level of reliance on machine outputs.

We argue that there needs to be different types of transparency and explainability for different purposes. For example, transparency for the purpose of problem diagnosis requires knowing why the system is exhibiting undesired behaviour in order to remedy it (Haugh et al., 2018). Transparency for the purpose of prediction in order to field a weapon requires being able to forecast how the system will behave in given circumstances. (Haugh et al., 2018). Transparency and explainability for the purpose of bounding a system requires understanding the limits of dependable performance in order to formulate tactics, techniques, and procedures for using the system, as well as identifying when monitoring the state of a system during its operation may be the only way to avoid undesirable behaviour.

One way to assure the type of transparency that TEVV might require is to have systems for recording metadata. Wojton, et al. for example suggest that '[i]f systems are recording data about their own decisions and internal processing, then stakeholders, including developers, testers, and even users, can gain more transparency into the system' (2021, p. 20). With respect to TEVV, this might be combined with 'safety middleware or disabled functionality to execute what some call 'shadow testing,' where the complex system makes decisions about what it would do in the current situation without being allowed to implement or execute those actions' (2021, p. 20). Such shadow testing could also provide meaningful updated data from the operational environment as well as the equivalent of counterfactual explanations for certain behaviours that can be useful for the TEVV process and operators.

TEVV Should Define the Appropriate Operating Environment

TEVV is meant to provide insights into variable performance in a range of operational environments (the so-called operating envelope) for all systems, whether conventional or AI-enabled. When it comes to AEW, however, defining the operating envelope is significantly more complicated. TEVV therefore has a more significant role to play in defining appropriate operational environments for the use of an AEW than for conventional weapon.

TEVV can also provide some insight into the initial decision whether a particular algorithm should be used instead of a human or non-AI alternative. There may be times when an AEW that works relatively well in a given context will not do so better than a human. This means that TEVV ought not to assess the safety and precision of a weapon in a vacuum, but with an eye to the comparative benefits and risks of reliance on humans or machines for similar functions in varied operational environments.

TEVV Should Drive Certification Schemes for AEW Operators

The iterative process used in TEVV can help guide appropriate training, skills and certifications for operators of AEW. For example, the US Joint AI Centre proposed including four types of testing: algorithmic testing, human-machine testing, systems integration testing, and operational testing with real users in realistic scenarios. The human-machine testing and the operational testing provide evidence not just for the evaluation of the weapon, but for how best a weapon should incorporate and present machine outputs in order to augment human judgment in the decision-making process. While TEVV has always played a role in US certification schemes for operators, the training content that can emerge from TEVV of AEW may well be significantly greater.

TEVV Should be Undertaken in Various Configurations of Systems and Humans

As described above, AEW are often systems of systems – that is, chains of algorithms with one algorithm's output serving as input for another. In such cases it may not be possible or desirable to test only one algorithm at a time. This suggests that ML algorithms will need to be tested in various configurations, operating with, alongside, and/or in a chain with several other ML models/algorithms. Similarly, some scholars have argued that rigorous testing should focus on testing various configurations of both systems and humans (Hand & Khan, 2020).

A TEVV process that is sensitive to the unique challenges of AEW can meet the requirements of the *jus ante bellum* by helping ensure that a State develops and fields these weapons in ways that minimize the distinctive risks that they can pose. As the next section discusses, however, this alone will be insufficient to meet these requirements if a State acts in ways that create the risk of a security dilemma.

AI-FUELLED SECURITY DILEMMA

A security dilemma exists when one State's investment in military capabilities prompts other States to increase their own investments because they perceive that the first States's actions make them less secure. Even if every States's investment is only for defensive purposes, uncertainty of other States about this intention can lead them to increase their own investment to protect themselves. This in turn can increase other States' perception of insecurity, resulting in escalation and increasing risk of violent conflict.

Two factors may be especially important in determining whether a security dilemma arises. The first is whether States perceive that the offense or the defence has the advantage. As Robert Jervis puts it, '[W]hen we say that the offense has the advantage, we simply mean that it is easier to destroy the others' [forces] and take its territory than it is to defend one's own' (Jervis 1978, p.-187). When a State believes this is the case, it is likely to conclude that it 'cannot afford to wait until there is unambiguous evidence that the other is building new weapons because the war may be over before it can get arms to its forces' (1978, p.189).

A second factor is the ease of distinguishing development of offensive and defensive weapons. The ability to differentiate between the two allows non-aggressive States 'to behave in ways that are clearly different from those of aggressors.' (Jervis, 1978, p. 199). In this situation, States can effectively signal their intentions by the type of weapons that they develop. One States's investment therefore need not make other States feel more vulnerable, reducing the likelihood of a security dilemma. When such differentiation is not feasible, however, States are likely to assume the worst and believe that they need to increase their own weapons investment, which reinforces a dilemma.

The risks and unpredictability of AEW may naturally lead States to refrain from fielding them until there is assurance of their safety and reliability. As several observers have suggested, however, various features of AI may increase the likelihood of a security dilemma that creates incentives for States to develop and field AEW soon as possible without a rigorous TEVV process to provide such assurance. (Altmann & Sauer, 2017; Ayoub & Payne, 2016; Bode & Huelss, 2018; Garcia, 2018; Horowitz, 2018; Payne, 2018).

Uncertainty about Capabilities

First, AEW will not be directly observable in the way that conventional weapons are. Whether a system is enabled by AI depends not upon its visible physical characteristics but the software that guides its operation. Two weapons that are identical in appearance therefore may have dramatically different capabilities. This means that it may be extremely difficult for one State to determine the AEW capabilities of another. This opacity surrounding AI capabilities could lead States to adopt worst case assumptions about the threat posed by other States with AEW.

Second, the dynamic rate of AI innovation means that even if it were possible to make an assessment of a State's AI-enabled capabilities at one point, this assessment may soon be outdated. States are likely to stay abreast of AI research and development, and to seek continuously to incorporate new capabilities into their systems. This 'uncertainty of measuring relative progress in AI research and its military applications' can make it difficult for States to have a stable understanding of the balance of power (Imbrie & Kania, 2019, p. 10).

Third, AI is not itself a weapon but a technology that can be used in various ways. As Horowitz et al. observe: 'As an enabling technology with many discrete applications, the amorphous quality of AI exacerbates uncertainty over how its integration into existing platforms and doctrine will change the character of warfare' (2020, p. 539). A State therefore faces a considerable challenge in attempting fully to comprehend all the ways in which other States may be incorporating AI into their military operations.

Fourth, at least in the near term, States have little experience with the use of AEW that could provide a shared understanding of their capabilities, limitations, and risks. This contrasts with conventional weapon systems and, notably, with nuclear weapons. An understanding of the devastating impacts of nuclear weapons provided a basis during the Cold War for the United States and the Soviet Union to take steps to reduce risk that such weapons would ever be used. There is not yet comparable clarity about the capabilities and risks of AEW that could lessen the likelihood of a security dilemma.

All these features of AEW systems are likely to mean that it will be extremely difficult to distinguish between offensive and defensive weapons. States therefore may be likely to believe that other States' development of such weapons threaten their security and that they need to develop and field AI enabled systems as soon as possible, including those for offensive purposes – which in turn will trigger other States' sense of insecurity.

Uncertainty about Intentions

Certain features of AI also may make it especially difficult for States to discern one another's intentions with regard to using the AEW that they have. First, the difficulty of distinguishing between offensive and defensive systems can make it difficult for a State to signal its benign intentions through its choice of the AEW in which it invests.

Second, there may be significant limits to how transparent States are willing to be about their AI-enabled capabilities, because transparency would involve disclosure of highly sensitive software. This software, rather than the platform in which it is used, could be a significant source of competitive advantage that a State would want to conceal. This creates challenges for any attempt to use inspection and verification as a way of reducing uncertainties and the risk of misperception about capabilities.

Perception of Acute Threat

Uncertainties about both capabilities and intentions with regard to AEW could lead States as a prudential matter to overestimate other States' capabilities and to assume that they have aggressive intentions. The nature of AEW may intensify this sense of threat because of the perceived decisive advantage of operating at machine speed compared to a 'remotely controlled, 'slower' adversarial system' (Altmann & Sauer, 2017, p. 119). A State may feel especially vulnerable because it fears that another States' use of such weapons against it would inflict such damage that it would be unable to respond. As Lieutenant General Jack Shanahan, the first Director of the Joint AI Centre, declared: 'What I don't want to see is a future where our potential adversaries have a fully AI-enabled force and we do not… I don't have the time luxury of hours or days to make decisions. It may be seconds and microseconds where A.I. can be used' (U.S. Defense Innovation Board, 2020, p. 3).

Under these circumstances, States are likely to believe that the balance of military capabilities favours the offense, which can make a preemptive strike seem advantageous. They therefore may hasten to develop and field offensive AEW in order to develop this capability. As Altmann & Sauer note, 'Destabilisation becomes a particular concern when qualitatively new technologies promising clear military advantages seem close at hand' (2017, p. 121). If 'the situation is seen as urgent… there are compelling incentives for accelerating the development of technology and incorporating it into militaries' (2017, p. 121). Furthermore, to the extent that use of an AEW involves less risk to military forces than a conventional weapon, a State's risk calculus may make a preemptive strike seem more appealing.

Finally, as Horowitz and Scharre suggest, a State's fear that AEW could quickly disable command and control capabilities could lead a State to develop weapons that automatically fire without human intervention upon warning of an impending attack (2021). Experience from the Cold War indicates the risk of such pre-programmed responses, when a Soviet nuclear technician determined that an apparent warning of an imminent nuclear attack was the product of a technological failure, and likely prevented nuclear war by not following protocols (Aksenov, 2013). AEW therefore could place pressure on mechanisms that are designed to control the risk of escalation.

At its core, the security dilemma reflects States' difficulties in determining the extent to which other States pose a military threat to them. The greater the uncertainty about other States' capabilities and intentions, the greater risk of this dilemma. This risk may be exacerbated by indistinguishability between offensive and defensive weapons, and by the perception that the balance of power favours offensive use of military capabilities. The perceived decisive advantage of operations conducted at machine speed beyond the ability of humans to respond can make the fear of State insecurity especially acute and foster the conviction that the balance of power favours the offense. The result may be that States believe it is necessary to field AEW as soon as possible without rigorous TEVV. Ethical development and fielding of AEW systems therefore requires that States engage in these processes in ways that minimize the likelihood of these risks.

AVOIDING THE SECURITY DILEMMA

What might States do to minimize the risk that developing and fielding AEW will generate a security dilemma that could result in risky fielding and use of such systems? One step is to avoid using language likely to trigger a sense of insecurity in other States. A second is to explore ways to reduce the uncertainties that give rise to the dilemma, and to adopt measures that can build trust among States about how such systems will be used.

The Language of Military Preparedness

On the one hand, a State needs to signal to other States that its military investments are intended to enable it to defend itself effectively if it is attacked, and to win any conflict that occurs. On the other hand, it needs to communicate that its investments are not designed to give its military such a major advantage that other States feel so threatened that they must escalate their own investment without regard for ethical concerns. Regarding the development of AEW, a State therefore should avoid characterizing its systems as providing it with an unprecedented decisive military advantage over other States. Given the potential of AI to conduct operations at machine speed, this may well trigger an intense sense of insecurity in other States. This insecurity could lead them to hasten development and fielding of AI systems without rigorous TEVV.

Language that can create the same risk is the public declaration that States are engaged in an 'AI arms race.' Unfortunately, there is no shortage of such language (Geist, 2016; Rickli, 2017). 'A 2019 survey of AI experts from technical and policy-oriented fields, for instance, indicated that an overwhelming majority of respondents predicted an AI arms race, however defined, in the next 15 years (Carpenter & Mahoney, 2019, p. 8). A risk of framing the situation in this way is that States may believe that they need to invest in developing and fielding AEW as soon as possible if they want to be secure (Scharre, 2021).

Just development of AEW thus requires that a State avoid language about such weapons that heightens States' insecurity. Aside from avoiding the use of such language, a State also may take affirmative steps to reduce other States' fears about its development of AEW. As the next section describes, one way to describe such steps is 'confidence building measures.'

Confidence-Building Measures

In the military context, confidence-building measures (CBM) are designed to reduce States' suspicion of one another through the exchange of information about capabilities and intentions, and to establish some agreement on how military operations will be conducted (Desjardin, 2014). Marie-France Desjardin's study of CBMs concludes that '[i]ncreasing transparency in military matters lies at the core of the confidence building approach… Secrecy breeds suspicions, and when States do not communicate, or there is a lack of information about other States' military capabilities or activities, officials tend to make worst-case analyses' (2014, p. 21). There is a rich body of work on the potential of such measures to help reduce uncertainty about State capabilities and intentions that can fuel a security dilemma (Horowitz et al., 2020; Horowitz & Scharre, 2021; Imbrie & Kania, 2019). The discussion below highlights some measures that observers suggest could accomplish this with respect to AEW.

CBMs gained particular prominence during the Cold War as a way of reducing the likelihood that misinterpretation of capabilities and intentions could lead to nuclear war. One example was the Open Skies agreement, under which the United States and the Soviet Union agreed to permit aerial surveillance to establish their missile capabilities. Another was the Incidents at Sea Agreement that regulated the movement of US and Soviet naval vessels, and established means to communicate the presence of submarines and surface naval movements. A third example is the creation of a hotline for communication between top US and Soviet Union leadership after the Cuban Missile Crisis. In addition, the 1972 Anti-Ballistic Missile Treaty imposed certain limitations on nuclear weapon capabilities. Finally, NATO and the Soviet Union agreed to notify one another of military exercises above a certain threshold to reduce the risks of escalatory responses.

It is important to appreciate, however, that the conditions regarding AEW may not be completely comparable to those that provided incentives for CBMs during the Cold War.

First, both during the Cold War sides were keenly aware of the destructive power of nuclear weapons because of the bombs that had devastated Hiroshima and Nagasaki. This created a common interest in avoiding mutual annihilation. By contrast, the impacts and risks of AEW are unclear at this point because they have not been widely fielded or used, particularly for offensive purposes. Second, weapons' capabilities during the Cold War generally were discernible from physical observation. This made it possible to use surveillance to assess such capabilities, and to engage in verification of compliance with arms control agreements. By contrast, as described above, the capabilities of AEW are not readily observable, but are in software that States may be reluctant to disclose. This opacity can give rise to a third challenge: it may be difficult to verify whether harm caused by an AI-enabled system was accidental or intentional.

Finally, States were the parties who engaged in the development of nuclear weapons during the Cold War, and they relied on highly centralized systems to control their use. By contrast, the developments in AI are generated to a significant degree in the private sector, the technology is widely available, and the uses to which it can be put are manifold. As Altmann and Sauer suggest,

> While the development of AWS [automated weapon systems] clearly presents a challenge to less technologically advanced actors, obtaining AWS with some degree of military capability is a feasible goal for any country already developing, for example, remotely controlled armed UAVs.
>
> (2017, p. 126)

These features of AEW suggest caution in assuming that we can rely on approaches during the Cold War to reduce risks from the use of these systems.

With these considerations in mind, there may be some steps that a State could take to reduce the concerns that underlie the security dilemma, and thereby engage in ethically responsible development of AEW systems.

One step is for a State to officially announce that it is committed to ensuring that fielding these systems is consistent with ethical principles and legal requirements, and that there is assurance of their reliability and safety. The US Department of Defense (DoD) (2023), for instance, has adopted ethical principles for the use of AI (DoD, 2022). DoD has followed this with guidance on how to implement these principles across the Department through the continuous identification, evaluation, and mitigation of risks (DoD, 2022). Similarly, DoD is adapting both its TEVV and weapons review process to conduct assessments of AEW. These steps will involve additional financial costs, as well as some delay in fielding systems. Commitment to them therefore can serve as a 'costly signal' to other States that they will not be disadvantaged by likewise fielding AEW only after rigorous review (Salahshou, 2019).

A second step could be to work to develop common definitions and shared understanding among States of core concepts that are relevant to the safety, reliability, impact, performance, and risks of AEW. As Horowitz, Kahn & Mahoney suggest, 'Knowledge about the material facts, that is, the technical and organizational features of military-relevant AI applications, is the foundation on which to build understandings about the risks themselves and the means to reduce them' (2020 p. 541). Differences in how concepts are described can reflect differences in how they are understood, which can increase the likelihood of misunderstandings that could lead to tensions.

A third measure could be to encourage information sharing and the development of communication channels among States. Some degree of transparency about TEVV, for instance, could involve public release of general information about the process for assessment of AEW without disclosing their specific technical features. This would be similar to the US approach to weapons review, which involves disclosing the process but not the review of particular weapons, in an effort to encourage other States to conduct reviews. States might also share information on how to establish parameters that limit the domain in which a system can operate without human supervision, and how safely to shut it down if it begins to pose risks by operating beyond that domain.

There could be some risk to a State from sharing such information, since it may enhance the ability of adversaries to field effective and reliable systems that they could use to threaten the sharing State's security. As Imbrie and Kania put it,

> On the one hand, collaboration in AI safety and security can reduce the risks of accident and strategic miscalculations among great powers. On the other hand, such collaboration may improve the reliability of machine learning (ML) techniques and therefore enable strategic competitors to field AI/ML-enabled military systems more quickly and effectively.
>
> (2019, p. 4)

A State therefore would need to decide how to weigh the security risk of an adversary's improved AI capabilities compared to the risk of an adversary and other States fielding AEW in ethically problematic ways. One way to address this could be encouraging ongoing exchanges among technical experts, members of the private sector, and academics, which could provide informal channels for sharing information that do not require official State involvement.

The measures described above could also help build confidence by serving as the impetus for a fourth step, which is establishing common norms and codes of conduct about the fielding and use of AEW. The United States, for instance, has promulgated a *Political Declaration on Responsible Military Use of Artificial Intelligence and Autonomy*, which has been endorsed by more than 50 States (U.S. Department of State, 2013). Among other things, the Declaration says that States should ensure that military AI capabilities are subject to 'rigorous testing and assurance' and that there are 'appropriate safeguards' to 'detect and avoid unintended consequences' of these systems.

Finally, States might work to develop 'rules of the road' for the use of AEW and perhaps 'red lines' that establish limits on their use. Horowitz and Scharre, for instance, suggest that an International Autonomous Incidents Agreement, similar to the Incidents at Sea Agreement, could provide rules to govern and deconflict the interaction of military forces operating with a high degree of autonomy (Horowitz & Scharre, 2021, p. 16). States also could agree to declare some geographic areas off limits to autonomous systems because of their risk of unanticipated interactions. 'This could be to avoid unintended escalation in a contested region (*e.g.*, a demilitarized zone), or because a region is near civilian objects (*e.g.*, a commercial airliner flight path)' (2021, p. 18). In addition, States might agree that AI systems should not be used for crucial functions related to nuclear weapons (Johnson, 2020; Maas, 2019).

CONCLUSION

Jus ante bellum requires that States conduct rigorous TEVV of AEW before they field them. It also requires that States develop and field AEW in ways that minimize the risk of a security dilemma that leads other States not to do so. States attempting to satisfy this requirement will need to balance competing ethical considerations. On the one hand, sharing some information about AEW may reduce other States' sense of insecurity and lead to measures that could enhance cooperation and minimize risks from these weapons. On the other hand, a State has an ethical responsibility to protect its population. It therefore will want to describe its capabilities in a way that discourages attack, and will have to assess what level of transparency about its AEW is consistent with ensuring its own security.

There is no formula to guide States in how best to navigate this tension. The important point, however, is that *jus ante bellum* insists that they attempt to do so in good faith in order to engage ethically in preparing for the possibility of war. It requires, in other words, that they focus on 'how to foster responsible competition' with regard to AI-enabled weapons (Roff, 2019).

REFERENCES

Aksenov, P. (2013). Stanislav Petrov: The man who may have saved the world. *BBC News*, 26 September. https://www.bbc.com/news/world-europe-24280831

Altmann, J., & Sauer, F. (2017). Autonomous weapon systems and strategic stability. *Survival*, *59*(5), 117–142. DOI: 10.1080/00396338.2017.1375263

Ayoub, K., & Payne, K. (2016). Strategy in the age of artificial intelligence. *Journal of Strategic Studies*, 39(5–6), 793–819, DOI: 10.1080/01402390.2015.1088838

Bode, I., & Huelss, H. (2018). Autonomous weapons systems and changing norms in international relations. *Review of International Studies*, *44*(3), 393–413. DOI: 10.1017/S0260210517000614

Carpenter, C., & Mahoney, C. (2019). How emerging technologies are rewiring the global order: Fall 2019 colloquium report. University of Pennsylvania Perry World House. https://www.academia.edu/45185410/Global_Order_Colloquium_Report_How_Emerging_Technologies_Are_Rewiring_the_Global_Order

Desjardins, M.-F. (2014). *Rethinking confidence-building measures*. Routledge. DOI: 10.4324/9781315000459

Flournoy, M. A., Haines, A., & Chefitz, G. (2020). Building trust through testing: Adapting DOD's test & evaluation, validation & verification (TEVV) enterprise for machine learning systems, including deep learning systems. Center for Security and Emerging Technology. https://cset.georgetown.edu/wp-content/uploads/Building-Trust-Through-Testing.pdf

Garcia, D. (2018). Lethal artificial intelligence and change: The future of international peace and security. *International Studies Review. 20*, 334–341.

Geist, E. M. (2016). It's already too late to stop the AI arms race—We must manage it instead. *Bulletin of the Atomic Scientists*, *72*(5).

Hand, D. J., & Khan, S. (2020). Validating and verifying AI systems. *Patterns*, *1*(3). https://pubmed.ncbi.nlm.nih.gov/33205105/

Haugh, B., Sparrow, D., & Tate, D. (2018). The status of test, evaluation, verification, and validation (TEV&V) of autonomous systems. Institute for Defense Analysis. https://www.ida.org/-/media/feature/publications/t/th/the-status-of-test-evaluation-verification-and-validation-of-autonomous-systems/p-9292.ashx

Horowitz, M. (2018). Artificial intelligence, international competition, and the balance of power. *Texas National Security Review*. 1(3), 37–57.

Horowitz, M., & Scharre, P. (2021). AI and international stability: Risks and confidence-building measures. Center for a New American Security. http://www.jstor.org/stable/resrep28649

Horowitz, M. C., Kahn, L., & Mahoney, C. (2020). The future of military applications of artificial intelligence: A role for confidence-building measures? *Foreign Policy Research Institute*. Fall, 528–543.

Imbrie, A., & Kania, E. (2019). *AI safety, security, and stability among great powers*. Center for Security and Emerging Technology.

Jervis, R. (1978). Cooperation under the security dilemma. *World Politics*, *30*(2), 167–214.

Johnson, J. (2020). Deterrence in the age of artificial intelligence & autonomy: A paradigm shift in nuclear deterrence theory and practice? *Defense & Security Analysis*, *36*(4), 422–448, DOI: 10.1080/14751798.2020.1857911

Maas, M. M. (2019). International law does not compute: Artificial intelligence and the development, displacement or destruction of the global legal order. *Melbourne Journal of International Law*, *20*(1), 29–57. DOI: 10.3316/informit.20190903016408

Payne, K. (2018). Artificial intelligence: A revolution in strategic affairs? *Survival*, *60*(5), pp. 7–32 DOI 10.1080/10.1080/00396338.2018.1518374

Pinelis, J. (2022). Progress in TEVV, SERC Talk. https://www.youtube.com/watch?v=1eSKngsJvvo

Rickli, J. (2017). Artificial intelligence and the future of warfare. *World Economic Forum, The Global Risks Report*. https://www.weforum.org/reports/the-global-risks-report-2017

Roff, H., & Danks, D. (2018). Trust but verify; The difficulty of trusting autonomous weapons systems. *Journal of Military Ethics*, *17*(1). 2–20, DOI: 10.1080/15027570.2018.148190

Roff, H. M. (2019). The frame problem: The AI "arms race" isn't one. *Bulletin of the Atomic Scientists*. DOI: 10.1080/00963402.2019.1604836

Salahshou, M. (2019). Evolution of costly signaling and partial cooperation. *Scientific Reports*, *9*, 8792.

Scharre, P. (2018). A million mistakes a second. *Foreign Policy*. https://foreignpolicy.com/2018/09/12/a-million-mistakes-a-second-future-of-war/

Scharre, P. (2021). Debunking the AI arms race theory. *Texas National Security Review*, *4*(3).

U.S. Defense Innovation Board. (2020). AI principles: Recommendations on the ethical use of artificial intelligence by the Department of Defense. US Department of Defense. https://www.ai.mil/blog_02_26_21-ai_ethics_principles-highlighting_the_progress_and_future_of_responsible_ai.html#:~:text=These%20principles%20encompass%20five%20areas,lifecycle%20both%20interactively%20and%20iteratively

U.S. Department of Defense. (2022). U.S. DoD responsible artificial intelligence strategy and implementation pathway. https://media.defense.gov/2022/Jun/22/2003022604/-1/-1/0/Department-of-Defense-Responsible-Artificial-Intelligence-Strategy-and-Implementation-Pathway.PDF

U.S. Department of Defense. (2023). DoD Directive 3000.09, Autonomy in Weapons Systems. https://www.esd.whs.mil/portals/54/documents/dd/issuances/dodd/300009p.pdf

U.S. Department of State, Political Declaration on Responsible Military Use of Artificial Intelligence and Autonomy (2013).

Van der Linden, H. (2010). Just military preparedness: A new category of just war theory. Paper presented at the Department of Philosophy at Michigan State University. https://digitalcommons.butler.edu/facsch_papers/1073

Vestner, T., & Rossi, A. (2021). Legal review of war algorithms. *USNWC*. https://digital-commons.usnwc.edu/ils/vol97/iss1/26/

Wojton, H., Porter, D., & Dennis, J. (2021). Test and evaluation of AI-enabled and autonomous systems: A literature review. Institute for Defense Analysis. https://testscience.org/wp-content/uploads/formidable/20/Autonomy-Lit-Review.pdf

2 Artificial Intelligence in Decisions on Armed Conflict

Robert H. Latiff

INTRODUCTION

The conduct of military operations by civilized nations is governed by the precepts of Just War Theory as embodied in the Laws of Armed Conflict and International Humanitarian Law. These sets of rules are general in nature and, while the types of weapons used in conflict evolve, are not technology dependent. New technologies have been developed and employed in warfare for centuries. However, technology notwithstanding, ultimately human beings' lives and livelihoods are the targets of war. In the present age, the militaries of advanced countries treat war largely as a contest of opposing technologies, weapons and equipment, but it is more than that. War is fundamentally a human endeavour. It is not, as the US Army points out, 'a mechanical process that can be controlled precisely, or even mostly, by machines, statistics, or laws that cover operations in carefully controlled and predictable environments. Fundamentally, all war is about changing human behaviour. It is both a contest of wills and a contest of intellect.' (U.S. Army, 2016, p.1-4). There are many components of military operations which in the past have been carried out by humans, but which are amenable to, and improved by, automation. There are, on the other hand, many components that, while they may be aided by automation, must remain under human control. The proponents of artificial intelligence (AI) such as Arkin (2007) argue that such systems can do any task a human does faster and more efficiently and argue that an AI can make better decisions than a human because it has access to enormously more information. Sceptics (such as Michel, 2023) argue that AI cannot substitute for humans in decision making about combat, about time-critical life and death situations and about actions requiring human emotions and judgement. In assessing the elements of Just War Theory and the factors to be considered in questions of *jus ad bellum* (justice in going to war) and *jus in bello* (justice in war), it will be clear that while artificial intelligence can be a useful tool, it has no legitimate role in actual decision making about questions of war.

Some fundamental basic concepts of military operations planning and execution in which automated systems can and do play a major role are described in some detail in this chapter. Within each major component of military planning and operations, there are numerous roles in which AI could be usefully and safely employed,

 DOI: 10.1201/9781003440253-2

and examples are provided. *Jus ad bellum* and *jus in bello* each have requirements that must be satisfied. Satisfaction of these requirements depends upon critical human judgement. In many of these requirements, there are roles in which proponents of AI believe that such systems could be substituted to make what were previously human decisions. This paper argues that, while AI can assist if properly employed, it is merely a tool and cannot substitute for humans in decisions about war.

MAJOR FUNCTIONS IN MILITARY PLANNING AND WHERE AI MIGHT HAVE A ROLE

Military planning and operations follow a well-established and historically successful process. The process is 'an iterative, logical planning method to select the best course of action for a given battlefield situation' (Meerveld, Lindelauf, Postma & Postma, 2023). Some military planners refer to this continuous and iterative sequence as the OODA Loop (Observe, Orient, Decide, Act) process.

The OODA Loop describes perceptual activities, decision making activities, and operational activities. Perceptual activities would include such things as intelligence analysis, situation assessment, threat detection, reconnaissance, and target recognition. In what is sometimes referred to as the intelligence preparation of the battlefield, the operator first collects as much information as possible about the enemy, the terrain, and any other external factors that might affect the operation and any activity that enhances the soldier's knowledge is employed. The commander will employ multiple technical sensor capabilities and intelligence sources at his or her disposal. In the observe phase AI has shown itself to be a useful tool. It will play an important role in gathering and analysing the large amounts of data collected.

The commander then assesses the status of his own forces vis-a-vis those of the enemy, creates plans and potential courses of action, assesses his or her unit's resource needs, decides then executes the plan. Calling upon experience and training, they orient their units to the realities of the battlefield they have just observed. They then decide on the best course of action based on past experiences and possibly a simulation of outcomes based on assumed enemy reactions. All of the military services have plans to incorporate AI into both administrative and logistics systems, and into command-and-control systems at every level in the chain of command.

The decision phase is highly human-cantered. In it, AI could be used as a tool to aid in weighing information, simulating scenarios and providing possible courses of action for the commander to decide upon, but commanders, and commanders alone, must decide about the use of force, especially lethal force, where, when, how much and with what weapons. In the action phase of a military operation, the commander issues orders and weapons are employed. Clearly there will need to be consideration here of the zone of battle (isolated battlefield or urban combat) and types of weapons to be employed. As adversaries engage on the battlefield, weapons, both physical and cyber, and possibly autonomous, are employed against the enemy and information from each phase is then incorporated in a continuous feedback loop, improving the intelligence, updating assumptions about the enemy, and determining the effects of friendly force actions. Again, such algorithms and simulations must be taken for what they are worth – based on assumptions and approximations of the current situation.

It should be noted here that there are different categories of AI systems. Often called AI are the so-called expert systems that utilize large amounts of generally well behaved data and incorporate machine coding of expert experience, processes and techniques. For military applications of expert systems, consider large logistics or transportation management systems where there is a plethora of past data experience. The other major category of AI system, and the one of most concern, are those that use machine learning, often employing neural networks or similar techniques. These also use large amounts of data, which is often not well behaved or curated, and attempt to find patterns or draw conclusions based on that data. In the case of combat, where each situation is unique and where data could be both chaotic and sparse, the use of AI and machine learning is problematic. Lamparth and Schneider (2024) note that the behaviour of large language models (LLM), especially in uncommon and unpredictable situations, can be erratic and, because no two are exactly alike, they are very sensitive to, and uniquely influenced by, user inputs.

AI, SMART WEAPONS AND THE THRESHOLD FOR VIOLENCE

Weapons and military systems employing various forms of AI are often referred to as smart weapons or smart technologies. Smart technologies are those that integrate computing and telecommunications. Smart devices are often networked with other similar devices and allow automated or even adaptive behaviour. Such devices and networks increasingly use artificial intelligence, machine learning, and big data analysis which, when combined with the ubiquitous sensors in an internet-of-things, provide something that attempts to mimic cognitive awareness.

In a military context, weapons employing smart technologies are computer-controlled and embedded with sensors and guidance systems. Some of the technologies employed in smart weapons include advanced robotics and digital technologies, including AI. Smart weapons can be missiles, bombs and precision artillery, to name a few. They can be, but are not necessarily, autonomous. While in recent years the major concerns about AI in military systems has centred on lethal autonomous weapons, we must now also include the rapidly growing use of computers, big data, and AI in military command and control and decision-making support systems.

In the context of just war theory, a key question about new technologies is whether the presence of smart weapons on the battlefield lowers the threshold for war and if so, what are the moral implications? One way to lower the barrier to violence is to reduce the risk to the attacker and one way to reduce the risk to the attacker is to increase the separation from the battlefield. Militaries have been doing this for centuries, from the introduction of the long bow in the fourteenth century, to artillery, to ballistic missiles and guided munitions such as drones and cruise missiles. Distance is important. Studies have shown that soldiers at close range or in hand-to-hand combat exhibit much higher resistance to killing. (Hughbank & Grossman, 2013, p.495). At long range, snipers, artillery, bombers, resistance is much lower. Technology facilitates distancing. It helps soldiers to overcome natural resistance and makes killing easier. Distancing and the resulting emotional disengagement can make warfare safer for soldiers but increases the potential for increased harm that accompanies them. Not actually seeing the target can lead soldiers to kill people and destroy property that they otherwise never would. Soldiers who see adversaries as mere icons on a screen or pilots who bomb from high altitudes have fewer inhibitions about their actions than soldiers in close combat. Distancing will lower the threshold for war. Another incentive to go to war might be reduced cost. Computers, software and AI systems are comparatively inexpensive and widely available. In addition, satellite imagery and communications are now commercially available. Precision guided munitions (PGM) have dramatically reduced collateral damage in war and lowered the number of non-combatant casualties. In the past, multiple 'dumb' bombs were needed for one target. With PGMs, we now have much higher accuracy. Even though they are more expensive, they require fewer weapons per target, resulting in a low cost-benefit ratio. They, too, provide for long range standoff from the battlefield. Sales and use of PGMs worldwide remain high and it could be argued that the increased use of such weapons is a direct result of their availability and ease of use. While PGMs decrease the chances of harm to non-combatants, they may increase the potential for their own use, thus skewing the ideas of proportionality or military necessity and lowering the barrier to violence.

JUST WAR THEORY AND WHERE AI MIGHT HAVE AN IMPACT

Jus in bello, justice in war, requires that military actions should be proportional, militarily necessary, and do not employ weapons that are considered inhumane. It is questionable whether a machine would have the ability to make, or even mimic, reasoned judgements about questions of proportionality, discrimination, and military necessity. A machine might be used to do a cost-benefit calculation or a comparison of destructive power to determine if a response was balanced, but likely could not judge whether an action's harm was less than the military advantage gained. On questions of military necessity, given a presumed bias toward winning, a machine might calculate that a village located between friendly forces and their objective presented an unknown threat and needed to be eliminated. Such a calculation might overlook factors a human commander would consider in deciding to commit forces to a potentially unnecessary fight.

One key concern about the use of autonomous weapons is about a machine's ability to decide who is, and who is not, a legitimate combatant. Questions about a machine's ability to discriminate combatants from non-combatants are well known and are valid, given the uncertainties involved. Even given greater sensing and computing capabilities, the ability of machines continues to be less than perfect. Clearly, in an era in which many conflicts are fought in cities and among the populace, this is a key question. However, in the (increasingly rare) case of a clearly defined battlefield such as seen recently in Ukraine, where all personnel in the zone of operations are considered combatants and the commander has made the decision and demonstrated intent to deploy such weapons, whether a weapon is autonomous or not seems to be a moot point.

While operational uses of AI in weapons and other battlefield uses fall under the category of *jus in bello* (justice in war), issues of command and control and associated decision-making fall under the area of *jus ad bellum*, or justice in going to war. If one thinks of preparation for, and decisions to undertake, war and combat operations, the questions that must be immediately answered are whether a proposed military action meets the criteria of just cause, right intention, proportionality, proper authority and last resort. The big question here is whether or not a machine is capable of making such decisions or if it would merely compute probabilities, absent any consideration of moral dimensions. Algorithmic systems output so-called 'decisions' after processing large amounts of data through a series of mathematical computations and statistical inferences. The problem is that having processed the data, the answers these systems give are very likely constrained by the narrow objectives for which the systems were designed, without regard, necessarily, for potentially harmful consequences that violate our moral standards of justice and fairness. (Moser, den Hond & Lindebaum, 2022).

How would a machine, for instance, calculate whether the cause for war is a just one? Would it somehow keep score of the preceding actions on each side, and assign each a weight? Is justice computable, or would a machine merely search the data for examples and try to find one that best fits the current situation? How could a machine determine of there was right intention? Would it search for public pronouncements of the nation's leaders to determine if war was a legitimate response? The concept of intentionality in a machine is nonsensical. Only human commanders can make such a call. Lucas (2013, p 218) states that 'machines cannot commit war crimes inasmuch as they lack intentionality or self-motivation and are utterly devoid of interests (or emotions).''Ethics and accountability,' he continues, 'remain firmly and solely in the domain of human experience.' Finally, while a machine can tally the pros and cons, only a human can be held accountable and thus can be the proper authority to initiate combat operations. Of course, as discussed earlier, once combat operations have commenced, we are faced with questions of proportionality in force application, distinction or protection of non-combatants, military necessity, and avoiding unnecessary and superfluous suffering. Each of these lends itself differently, or not at all, to mathematical calculations and none of them are the proper province of anything but a human decision maker.

Command and control is an increasingly important topic as big data and artificial intelligence find their way into decision support systems. Militaries plan to use machine intelligence to stitch many individual systems into a single network and while in the past, we discussed kill chains, now we refer to kill webs (Uppal, 2022). These webs link every weapon, commander, and soldier to every other in a system that will speed decision making and, presumably, provide more accurate, or at least more, data. In complex situations, under intense time pressures, it is unclear the extent to which humans can really exercise meaningful control.

To illustrate the concern about the high-speed, highly automated nature of current command and control, and possibly the decline of meaningful human intervention, consider the story of Lt. Gen. (retired) Hal Moore as he commanded US troops in the battle at Ia Drang, Vietnam in 1965 (Moore & Galloway, 2004). He described the intense combat and how he would sometimes step back slightly to gather his thoughts and reflect on what was happening and, as importantly, what was not. Moore's actions were uncommon and with today's style of warfare might even be less so. With too much information and too little time, leaders are deprived of the ability to think or reflect on the context of a situation. They do not have the luxury of what philosopher

Simone Weil calls 'between the impulse and the act, the tiny interval that is reflection' (Weil, 2007). Stewart and Hinds (2023) conclude that employing techniques such as tactical patience throughout the military decision making process offers advantages such as being able to see more, understand more and develop more options.

The digital world's emphasis on speed further inhibits reflection. Today's battles may not allow such a luxury. For instance, inbound hypersonic missiles give targets little time for decision-making and might force ill-informed and ill-advised counter-decisions. Cyber conflict occurs at speeds at which no human could keep up. Autonomous systems, connected in a network, create an efficient weapon system. A mistake by one, however, could race through the system with possibly catastrophic consequences. The speed of present day and potential future conflicts, coupled with the enormous complexity and volume of data ingested, make meaningful human judgement and control potentially impossible. The required speed of decision making can, and probably will, outpace our ability to exercise due care. Worse yet, military commanders are understandably predisposed to act, even when inaction might result in a better outcome. This would be even more of a problem in the face of urgent AI generated recommendations to do so.

If the speed of war were not enough of an issue, decision support systems themselves might contribute to lowering the barriers to war. Under pressure, humans have been shown to be more likely to defer to technology. Defence Advanced Research Agency programme manager, Matt Turek, noted that AI systems give the illusion of competency, enticing users to turn over tasks to them. (Michel, 2023). While soldiers can be instructed to treat information from these systems with appropriate scepticism, with machines that are most often right, there may be a reluctance to ignore the computer. What if the system was right and ignoring its recommendations created a failure that could have been avoided?

Among the dangers in the growing use of decision support systems are over-reliance, automation bias, and user complacency. Users of highly reliable automated systems build trust in machines. If systems fail in way that is not obvious, users fail to notice. If they are almost always right, there is difficulty keeping the expert actively involved throughout the process (Bolia, Vidulich & Nelson, 2006, p.8). There becomes an automation-induced complacency where systems suggest their own solutions and invite passive decision making on the part of the human (Cornelissen, van Eerdt, Schraffenberger & Haselager, 2022, p.1). Also, because the systems are so complex, the human operator may not recognize that available information is incomplete, or possibly wrong, and the user is unlikely to know either that they need it or where to look for it. The problem is compounded if we don't know how a model is wrong. As Moser et al. (2022) point out 'it is not only useless, it is dangerous.'

When humans encounter highly complex situations, our cognitive limitations often prompt us to outsource our decision-making to automated systems. Highly accurate automated systems tend to create automation bias in which human users give greater credence to information supplied by technology and ignore a manual source that provides contradictory information, even if correct. Users will select a path requiring the least cognitive effort, letting technology dictate. (Grissinger, 2019, p.321). Ultimately, technology's performance convinces the human that it is acceptable to hand over tasks to the system. This behaviour is more likely with complex tasks, multitasking, heavy workloads, or time pressures—all characteristics of a modern, active battlefield. Users rely on technology to inform them of a problem but if it does not, they fail to respond to a critical situation because they were not prompted. They become complacent and less vigilant because of lower suspicion of technology error, believing that technology's capability is superior to humans. Increased familiarity with technology can lead users to doubt their own instincts When they perceive automation as reliable most of the time, people are less likely to question its accuracy. Users have been shown to override their own correct decisions in favour of erroneous advice from technology. Worryingly, the technology-failure detection rate is also low. Users fail to detect any errors deliberately introduced during the course of a typical workday (Goddard, Roudsari &Wyatt, 2012, p.121).

Issues with AI Systems and Problems of Trust

There continue to exist some serious technical issues with AI systems and their trustworthiness. The unpredictable nature of AI systems and the inability of such systems to explain their reasoning processes are problematic for military commanders. A major issue is the data on which these systems learn. Real combat data is scarce and simulated data could, in fact, contain human bias. These systems and their data have also been shown to be very sensitive to spoofing and hacking and, since they are co complex, such spoofing could well be undetectable. AI systems have often been described as 'black boxes' since they cannot in any humanly recognizable way describe why and how they arrived at the decision or recommendation they provide. AI systems cannot be tested in the classic way in which military weapons are tested. Their behaviour is, by design, not deterministic. AI systems tend to be extremely complex and exhibit highly probabilistic behaviours. Evaluation is done with so-called end-to-end scenarios in which it is determined that given a reasonable input, the system returns a reasonable output. How it achieves the output is unknown.

The pervasive unpredictability of AI systems and computer thinking are different from human thought processes in often incomprehensible ways. Unpredictability is something human commanders try to mitigate. Finally, there is an understandable bias toward success in combat. Programming battlefield AI systems to win might cause ethics to take a back seat to the primary goal. As one author notes 'these systems lack human common sense as computers seek the most direct solution to problems, not the most ethical or practical one (Fryer-Biggs, 2021).'

AI systems may very well contribute to the escalation of conflicts. They increase the pace of warfare, reducing commanders' decision time. As mentioned above, they may result in automation complacency thus adding to the problem of humans being out of the loop. There could also be inadvertent escalation as AI systems increase ambiguity, possibly resulting in a misperception or an incomprehensible assessment of the situation.

In a direct analogy with the civilian world, the military operates with an internet-of-things to include space, airborne, and terrestrial sensors, wearable devices, munitions, weapons, vehicles and robots. Because these components are largely connected wirelessly, there will be an extremely dense electronic environment on the battlefield.

The situation will often be chaotic, and commanders will face increasingly ambiguous tactical situations. The insertion of AI and automated decision systems, while they may speed the action, will not necessarily serve to clarify or elucidate. A former U.K. intelligence official said that

> because technology has allowed humans to connect, interact, and share information almost instantaneously anywhere in the world, this has opened channels where misinformation, blurred lines, and ambiguity reign supreme.
>
> (Hinchliffe, 2018)

Faced with ambiguous information, people will most often opt for the lowest regret, more conservative, response. In an ambiguous military situation, commanders might reasonably opt to take action, rather than not take action.

In discussing smart weapons, we must keep in mind ethics concerns, whether they can be used safely and reliably, whether they risk escalating wars more quickly and whether they comply with international humanitarian law. The use of unproven weapons whose effects are unpredictable is ethically and legally wrong. There remain serious concerns about explainability, vulnerability, fragility and unexpected emergent behaviour of AI systems. Currently it is difficult to understand or explain how a system arrives at its output. These systems have been shown to be vulnerable to bias in training data and models. They also have been shown to fail due to unanticipated operating environments. They are fragile in the face of deceptive input or training data, and they can be spoofed by adversaries. AI systems have in some cases demonstrated emergent aggressive behaviour. Finally, testing and predictability of performance on the battlefield is not possible and presents an issue of trust for soldiers whose lives may depend on such systems. Lucas (2013, p219) states that 'deploying a patently unreliable system is surely reckless (and therefore morally culpable) and in extreme circumstances may be found explicitly to be criminally negligent.'

Advanced information-processing and AI systems enhance situational awareness and the decision-making capacity of human commanders and soldiers. However, there is a concern that the end result may be elimination of those commanders and soldiers from critical decisions altogether. AI is only a tool, but humans are overusing it and depending on it, perhaps even when it is wrong. A machine could never replace – nor should it – human judgement in war. It may well be that such dependence on automated decision support systems will lead to an abrogation of command responsibility and accountability.

To conclude, smart technologies now play, and will continue to play, a major role in conflicts. All things being considered, they will lower the threshold for violence. This has serious implications for Just War Theory and the Laws of Armed Conflict, as those systems' goals are to decrease the instances and harm of such conflicts.

INABILITY OF AI TO REPLACE HUMAN FUNCTIONS IN WAR

In such important decisions as war and combat, data and algorithms will not be enough. On the battlefield, emotions are critical. Humans make decisions by being part of society, having emotions and interacting with other humans and objects. Sensory experience and social interactions are critical components of our behaviour and every bit of information perceived is understood and processed in the context of everything a person has experienced. A system like the currently popular Large Language Models can search for and ingest massive amounts of extant data and information and package it for human consumption with blazing speed, but it is merely a snapshot. The human mind can call upon years, perhaps decades of experiences, emotions and subtleties of analogous situations and make nuanced judgements and decisions. Azarian (2022) concludes,

> machines compute, minds understand. The mind recognizes subtleties that a purely algorithmic intelligence would be blind to. It is conscious experience, presumably, which enables us to not just reason, but to reflect on reasoning.

Ethically intelligent behaviour may require much more than rationality. While it may theoretically be possible to programme some form of operationally moral behaviour—rules of engagement for instance—that is not enough. Soldiers have the ability to recognize and respond to situations with moral and ethical challenges and subtleties. Machines cannot. They cannot experience emotions, which are important elements of moral awareness. The ability to recognize and decide about moral dilemmas is required for ethical judgment. Even if machines can be shown to mimic some basic form of intelligence, they would not embody the qualities we identify with the warrior ethos. On a battlefield, soldiers exhibit such things as camaraderie, loyalty, trust and altruism. As Moser et al. (2022) point out, 'judgement relies not only on reasoning, but also on such capacities as imagination, reflection, examination, valuation, and empathy. Therefore, it has an intrinsic moral dimension.'

Scientists repeatedly make the point that to be truly successful, AI systems must interact with social systems, that is, humans. Philosophers refer to our current world as a techno-social one in which our technologies and our social interactions are intertwined, where technology has shaped human social practices, values, and institutions (Vallor, 2016, p.2). AI researchers conclude that autonomous software and biologically constructed machines will not be sufficient to generate true machine intelligence, saying that much of what we think we know is actually distributed in the social networks of which we are a part They conclude that autonomous software and machines will not be sufficient to generate true machine intelligence, saying, 'in ways we still do not understand, our social communities and interactions in language are essential for general intelligence' (Denning and Lewis, 2019). Indeed, what is war but social interaction on a massive and violent scale?

Placed in the context of life and death decisions, the ability of AI to replace, or more accurately, mimic the abilities of humans becomes a serious concern. Human commanders still need to make such decisions. Professional soldiers are highly trained, and commanders take their decision responsibilities seriously. While they will resist explicitly allowing machines to make important decisions, the characteristics—speed, complexity, ambiguity—of contemporary conflicts may at some point, overwhelm even the best commanders. This we must avoid.

SUMMARY

Military operations planning is a deliberate process with well-defined and accepted steps. As the tempo of modern conflicts increases, the goal is for friendly forces to ack quickly enough to get inside the adversary's decision timeline. Given the changing character of war, as it tends to include faster, more complex, more ambiguous situations, the capability of automating what can be safely automated is welcomed and irresistible. In this, AI has a legitimate role to play, offloading only that which can be safely offloaded from the humans, giving them more time to focus on those things that truly require human involvement.

The precepts of Just War Theory, both *jus ad bellum* and *jus in bello*, call for soldiers and commanders to exercise judgements about what constitutes proper behaviour. The incorporation of AI and decision support systems in combat, while it can help in providing the information needed for leaders to make such decisions, presents challenges. These include the potentially dangerous assumption that these systems will behave as intended and the worry that users will brush aside the known shortcomings and vulnerabilities. (Cole, et al., 2024, p.12). Of concern, too, is the possibility that such systems will become ubiquitous and slowly usurp the role of the battlefield commander.

Just War Theory, the Laws of Armed Conflict and International Humanitarian Law were developed before the era of high technology weapons, but their pertinence and validity remain as strong as ever. Their key concepts were developed in earlier ages in an attempt to limit the excesses of war and the human suffering it causes. They insist on actions and decisions to be made by the human participants in a conflict. Human judgement cannot be reduced to an analytical exercise. In matters of life and death, war and peace there can be no computationally derived answer. Cole et al. (2024, p.12) conclude that

> the assessment of conditions for the application of lethal force, the decision to put lives at stake in going to war, and the behavior of personnel and weapons in war are all human activities, still optimally controlled and carried out by human commanders.

REFERENCES

Arkin, R. C. (2007). *Governing lethal behavior: Embedding ethics in a hybrid deliberative/reactive robot architecture* (Report No. GIT-GVU-07-11). Atlanta. Georgia Institute of Technology.

Azarian, B. (2022, June 8). The mind is more than a machine. *Noema Magazine*. The Berggruen Institute.

Bolia, R. S., Vidulich, M. A., & Nelson, W. T. (2006). Unintended consequences of the network-centric decision making model: Considering the human operator. Human Effectiveness Directorate, Air Force Research Laboratory. Wright-Patterson AFB, OH.

Cole, A., Howard, D., Latiff, R., Lucas, G., Reichberg, G. M., & Kaushik, R. (2024). Artificial intelligence in military planning and operations. *PRIO Paper*. Peace Research Institute Oslo.

Cornelissen, N. A. J., van Eerdt, R. J. M., & Schraffenberger, H. K. (2022). Reflection machines: Increasing meaningful human control over decision support systems. *Ethics and Information Technology*, *24*(2). DOI:10.1007/s10676-022-09645-y

Denning, P. J., & Lewis, T. G. (2019). Intelligence may not be computable. *American Scientist, 107*(6), 346. DOI:10.1511/2019.107.6.346

Fryer-Biggs, Z. (2021, February 17). Can computer algorithms learn to fight wars ethically? *The Washington Post.*

Goddard, K., Roudsari, A., & Wyatt, J. C. (2012). Automation bias: A systematic review of frequency, effect mediators, and mitigators. *Journal of the American Medical Informatics Association, 19*(1), 121–127. doi: 10.1136/amiajnl-2011-000089

Grissinger, M. (2019). Understanding human over-reliance on technology. *Pharmacy and Therapeutics, 44*(6), 320–375.

Hinchliffe, T. (2018, December 5). In an era of AI, you need human intelligence: UK spy chief. *The Sociable.*

Hughbank, R. J., & Grossman, D. (2013). The challenge of getting men to kill: A view from military science. In D. P. Fry (Ed.), *War, peace, and human nature: The convergence of evolutionary and cultural views* (pp. 495–513).. New York: Oxford University Press. DOI: 10.1093/acprof:oso/9780199858996.003.0025

Lamparth, M., & Schneider, J. (2024, April 9). Why the military can't trust AI: Large language models can make bad decisions—and could trigger nuclear war. *Foreign Affairs.*

Lucas, G. (2013). Engineering, ethics and industry. In B. J. Strawser (Ed.), *Killing by remote control* (pp. 211–228). New York: Oxford University Press.

Meerveld, H. W., Lindelauf, R. H. A., Postma, E. O., & Postma, M. (2023). The irresponsibility of not using AI in the military. *Ethics in Information Technology, 25*(14). DOI: 10.1007/s10676-023-09683-0

Michel, A. H. (2023, August 16). Inside the messy ethics of making war with machines. *MIT Technology Review.*

Moore, H. G., & Galloway, J. L. (2004). *We were soldiers once and young: Ia Drang - the battle that changed the war in Vietnam.* Novato, CA: Presidio Press.

Moser, C., den Hond, F., & Lindebaum, D. (2022, February 7). What humans lose when we let AI decide: Why you should start worrying about artificial intelligence now. *MIT Sloan Management Review Magazine.*

Stewart, R., & Hinds, G. (2023). Humanitarian law and policy, Algorithms of war: The use of artificial intelligence in decision making in armed conflict. International Committee of the Red Cross.

U.S. Department of the Army. (2016). *Army Doctrine Publication: Operations (Publication No. 3-0, p.1-4).*

Uppal, R. (2022, March 10). DARPA ACK will assist military commanders in making rapid tactical decisions in multi-domain battlefield. *International Defense, Security and Technology.*

Vallor, S. (2016). *Technology and the virtues: A philosophical guide to a future worth wanting.* New York: Oxford University Press.

Weil, S. (2007). The Iliad, or the Poem of Force. In B. Lawrence & A. Karim (Eds.), *On violence: A reader* (pp. xxx–xxx). New York: Duke University Press. DOI: 10.1515/9780822390169

3 *Jus in bello*
AI Is Not Just Another Weapon

Malcolm Brown, James McEvoy, and Eamonn O'Neill

Just War Theory has deep roots, widespread throughout the world and through history, and addresses both practical and ethical concerns. It is likely that, from the earliest manifestations of violent conflict, reflective humans will have considered ways to limit their, and their fellows', destructive potential. Later, these reflections will have been codified in different cultures. Paul Ramsey thought of Just War Theory as inherent in the idea of good politics itself (Ramsey, 1961, p. 86). Here, we draw specifically upon the tradition of Just War Theory as it has evolved within the Western, and broadly Christian, context since Augustine.

Artificial Intelligence (AI) has much shallower roots, counted in just a few years rather than centuries, but has already achieved global reach and importance. Like Just War Theory, AI raises practical and ethical concerns, the latter increasingly debated in applications from medicine (Dalton-Brown, 2020; Gichoya et al., 2023; Gundersen & Bærøe, 2022) to criminal justice (Bagaric et al., 2022; Miller, 2022). Where AI is applied in the conduct of war, it is inevitable that this broader debate on the uses and ethics of AI will inform and be informed by Just War Theory.

Throughout its history, Just War Theory has often distinguished between *jus ad bellum* – the conditions under which parties may legitimately go to war – and *jus in bello* – *how* such parties should behave in the conduct of war. While AI, with its sheer breadth of potential applications, no doubt has implications for both, we confine our scope to the latter. *Jus in bello* is concerned primarily with discrimination and proportionality in how combatants behave (Reed 2004, p. 42). Discrimination is about avoiding, as far as possible, direct or indirect harm to non-combatants. Proportionality is about avoiding needless destruction in pursuing ends that may be justified. Given the premise that AI can replace – to some degree and in some circumstances – the human as combatant and decision maker, we consider some of the ways in which this is likely to happen and how Just War Theory might affect the judgements that are made in consequence.

DOI: 10.1201/9781003440253-3

JUST WAR THEORY AND THE DEVELOPMENT OF AI

Although Just War Theory has ancient origins, it is often associated with the Christian tradition, starting with St Augustine (Langan, 1984) and being developed into its well-known form by Aquinas (*Summa Theologiae* Pt.II, section 2, q.40). This cultural and religious specificity raises important questions when conflicts develop between nations and cultures which do not share those traditions. Such questions intensify when we turn to the application of AI and related technologies where the deep consideration of ethical issues through the lens of cultural specificities is only just beginning. The religious origins of the concept of Just War are, for instance, thrown into sharp relief when a conflict emerges between a totalitarian regime and a culture framed by a Christian heritage. To use the words of William Temple from the period of the Second World War,

> If we start without the Christian pre-suppositions, the State or the community for which it acts will appear as a whole of which the citizens are merely parts. If the citizen is a mere part, his interest is properly subordinate to that of the whole. Moreover, inasmuch as *ex hypothesi* there is no God for the State to serve, the State becomes an absolute. Sovereignty is then its essential attribute – Sovereignty over its subjects, Sovereignty over against other States. ... So we arrive at the maxim 'The State is Power.' If this be granted, the method of exercising this power is a matter of expediency only. ... Concentration camps ... are then perfectly legitimate.
>
> (Temple 1941, p. 33)

Other theisms may posit a similar relationship between the individual and the state in contrast to the principle of totalitarianism and the Power State. But, even in a state with no formal religious foundation, the distinction between placing the salience of the individual citizen above or below the state differentiates a liberal democracy from totalitarianism. There may be religious origins behind that distinction but the distinction itself, rather than the religious root, is key to the applicability of some form of Just War Theory.

Jus in bello revolves around two moral considerations: proportionality and discrimination (Reed 2004, p. 42). While the tradition of *jus ad bellum* seeks to prioritise peace by constraining the conditions under which a nation may go to war, *jus in bello* also has its eye on peace by ensuring that a war justly entered into does not create the conditions for future wars through conduct which turns victory into vanquishment or leaves a wasteland in its wake. Thus, *jus ad bellum, jus in bello* and ultimately *jus post bellum* connect with one another to form an approach to conflict lying firmly within a framework of minimising harm and promoting a just resolution. While the whole tradition is clear that war is never a good in itself, the principles of proportionality and discrimination introduce elements of virtue including, for instance, the virtue of prudence, into the conduct of war. However, the notion of virtue, crucial to the thinking of Aristotle or Aquinas, has only relatively recently enjoyed a renaissance in the study of ethics. Particularly in the liberal cultural traditions of the West, everyday life is conducted more according to the canons of utilitarianism and impersonal mechanisms like the market that are not dependent upon the specifics of tradition.

Modernity has relied heavily on the market as an impersonal and culturally neutral mechanism to manage the controversies generated by plurality, and it is at the interface between the avowedly amoral market (*vide*. Hayek, 1944) and the oral traditions of Just War Theory that some of the keenest ethical questions about AI arise. Once a weapon evolves beyond the function of a tool controlled by a human mind and begins to take on some of the decision making capacities of the human, it shifts from being a commodity to having an element of moral agency in its operations. This is different from any weapon that has been produced before; now the tools themselves can have at least a share in decision making. It becomes crucial to ask how far the market mechanism which drives most R&D, production and distribution is capable of recognising the moral demands of Just War Theory and reflecting those principles.

As AI's capacity to take autonomous decisions grows, questions arise about the cultural values and principles that inform it. This raises the question of how Just War ethics can be incorporated into military technologies if the ethical assumptions of their designers and developers cannot be assumed. 'Military tradition' includes embedded approaches to the ethics of war and combat. The traditions of different armies are often ethically distinct, reflecting differing political and cultural ideologies. International approaches to what is, and is not, permissible in war – including the definition of war crimes – demonstrate that the ethics of war are contentious and hard to police. The Geneva Conventions have been perhaps the strongest area of international law covering *jus in bello*. The United Nations Charter (Chapter 7) allows the UN Security Council to determine where threats to peace or acts of aggression have occurred and to intervene to restore peace and security. But neither has secured universal acceptance or has been straightforward to enforce, demonstrating how the ethics of war can vary. Given this diversity, it is worth considering how the ethical assumptions built into AI design and deployment for military purposes might be developed in a manner consistent with the ethical traditions of the military institutions deploying them.

AI systems can incorporate all manner of biases, some by design and some unintended. Designed bias may be incorporated for different reasons – perhaps to add a control factor or to ensure that specific intended outcomes are realised. Indeed, we are arguing here for some forms of intended bias to be incorporated in military AI systems to give those systems a preferential bias to the ethics of the force deploying the application. For example, the principle of discrimination drawn from Just War Theory would require the design of AI which, in relevant circumstances, discriminates between combatants and non-combatants, just as human soldiers are trained and mandated to do in a Just War culture.

But AI systems can also incorporate *unintended* biases based on factors from the data used to train the AI (Shahbazi et al., 2023) to the biases of their human designers and implementers (Flores et al., 2024). Such biases often have cultural or ideological foundations (Peters & Carman, 2024) which may militate against the principles of Just War Theory if the ethical context is not fully examined. The question therefore arises: whose assumptions, prejudices and ethical principles are incorporated (wittingly or otherwise) in AI applications which make life and death decisions and which, in a military context, could potentially cross the line into war crimes? While human combatants can, and sometimes do, make bad decisions which breach the ethics of war, there is a danger that remitting decisions to AI will undermine the values that characterise one's own ethics of waging war but without the accountability demanded of human agents.

Can the AI designers and implementers in the current AI market and culture, with their own intended and unintended biases but without necessarily sharing the traditions and values inculcated into military personnel, develop systems which are culturally sensitive to what their military users understand their roles to be? A particular danger may lie in developing applications which operate with an overly simple utilitarian calculus of costs and benefits, rather than seeking to retain as much of a humane ethic as is possible in a conflict. The risk of buying into crudely utilitarian principles by default suggests that the design and development of AI for military uses should be strongly influenced at all stages by service personnel experienced in reflecting upon their own ethical military tradition and not simply outsourced to commercial developers and the market. This might seem to be just another, albeit contemporary, instance of the long recognised, if not always achieved, imperative to understand and meet user requirements in software engineering; a term itself popularised by NATO through its Science Committee conferences in 1968 and 1969. But, as we argue below, AI is not just another software product.

This is where an approach from virtue ethics can throw light on the problem. As MacIntyre (1982, p. 188) has argued, if we approach questions of ethical conduct by examining the nature of our practices (and the pursuit of war according to Just War Theory is a practice) it helps to distinguish the external goods of a practice from the internal goods. As illustration, he cites the game of chess which, after all, is only a stylised simulacrum of warfare. The external good of chess – the primary object – is to win the game. But one may train a person to win at chess by cheating and, in the process, destroy the internal goods of chess which can be summed up as playing chess well. Moving from chess to actual warfare, it may sound jarring to talk of aiming to 'win a war well' but that is what *jus in bello* implies. War fought in a way which is mindful of the principles of proportionality and discrimination may avoid offending the notion of a Just War and be a well fought fight.

If the system driving the application is designed with the objective of winning, it becomes crucial to know how 'winning' is understood. A nation goes to war with a set of objectives which can evolve under pressure of events or be lost sight of in the struggle for survival. The means by which objectives are pursued include the ability to incapacitate the enemy, often by use of force against their assets and personnel. If the design of systems focuses only on achieving limited means more effectively, the design parameters may ignore the war aims and focus instead on other measures of 'success' – perhaps in measurements such as the body count or the ability to lay waste enemy territory. Without a grasp of the war aims and the internal goods of the practice of war, reflected in the principles of proportionality and discrimination, a focus on efficiency in delivering the means of conflict could form the sole measure of a successful AI application on the battlefield. Not unnaturally, proportionality and discrimination are not the first considerations to come to mind when the security of a people and the survival of a nation and its lifestyle are in jeopardy. That is why they have to be the subject of a deliberate policy or rule. Consideration of the ethical content of AI applications in war needs to start at the requirements and design stages, long before the product is marketed to potential belligerents.

In particular, building in proportionality and discrimination may need to be related back to the idea of the individual's place vis-à-vis the state as articulated by Temple. When a new and contentious development arises in the technology of warfare, the argument may be made that 'we' must take this step because if we don't the other side will. Often, this argument carries justifiable weight. But the principles of *jus in bello* are, at least in part, about ensuring that 'we' remain recognisably the same 'we' after a conflict as before it. Who 'we' are is essentially about moral identity. As Temple suggested, one of the fundamental differences between regimes is that some make the individual subservient to the interest of the state (the 'Power State') and can thereby justify atrocities, whereas those who locate the state in the service of the citizen cannot be ruled by expediency alone. If the defence of the latter kind of state is achieved by treating every means as justified by the end of victory, the moral distinction between the two kinds of state largely disappears. No one – and no culture – is ever quite the same after a war but the fundamental moral character of a nation can endure. Indeed, the rhetoric of wartime leaders may appeal precisely to that character as a reason for fighting. If national moral character is, in part, formed in and through the Just War tradition, then the principles of *jus in bello* should be built, as far as possible, into the means by which such a nation wages war, including AI applications.

FROM HUMAN TO AI DECISION MAKING

AI's ability to outstrip human decision making, whether in terms of speed or the volume of data that can be digested, also opens the possibility of AI enhancing the ability of combatants to engage in conflict within the traditions of Just War.

The concept of the OODA loop (Observe, Orient, Decide, Act) was initially developed in the 1970s by Col John Boyd of the US Air Force (Richards, 2020, p. 143). He predicted that the ability to move around this loop quickly in the face of rapidly incoming, partial and conflicting information would become the defining feature of effective warfare. One must first Observe what has happened, Orient to the changing situation and understand it, Decide the right course of action and then Act, implementing that decision. The cycle then begins again. The aim is to move around the cycle faster than your opponent; if you can achieve this then you are the one setting the conditions to which the enemy is forced to react.

AI can move through this cycle far faster than any human can. While an individual human may move through this cycle in seconds, the higher up the decision making chain one goes, the more complex the problems become and the more time it takes. At Brigade level, where one individual cannot know the details of all aspects of communications, logistics, medical, enemy tactics and weapons ranges (to name but a few relevant factors), teams of Subject Matter Experts work together and it is not uncommon for the development of a plan to act to take around twelve hours. An AI system that can hold all that information and more and does not experience human delays (such as discussions and briefings) but can run computations at great speed, can give a solution in seconds. With that in mind, it seems very likely that AI is coming to warfare (MoD, 2022, p. 11).

The ability of AI to support the planning cycle may seem shocking if we contemplate handing complete control to AI to run battles (Woodcock, 2024). It is important, therefore, to retain the concept of the 'human in the loop' or at least 'on the loop', where a human must either remain an integral part of the decision making process or, at least, have the final say before a decision is enacted. For example, driving a car with satnav is very much a human in the loop. If the car were fully autonomous and the car presented route options to the human before self-driving, then we may say the human is on the loop. The key claim in many applications of AI is that it is there to support, not totally replace, the human decision maker; plotting routes, for example, or finding the best locations for artillery platforms or locating the best viewpoints for observation posts. A human would still have the final say but the OODA loop has been made faster.

Can we envision a time where the human is removed entirely from the process of acting, perhaps lethally, in a changing environment without violating the concepts of Just War? We may already be there in some specific scenarios. The CIWS (Close-In Weapon System) is a defensive system typically found on ships and on land to protect military bases. It is essentially the last line of defence, being a point-defence weapon system that has a role of shooting down missiles, shells and other objects that pose a threat. While it could be manually operated, the speed at which its targets are moving makes this less than optimal, and so some systems such as the Phalanx are automated. This system saves lives precisely because it can do what a human could not. There are, therefore, some occasions when it is not only convenient or more efficient to hand over decision making and control to a machine but is a necessity. Comparing this example with the Brigade-level instance of the OODA loop, a key difference may be that CIWS is straightforwardly defensive against a speed of attack that human decision making cannot match. The Brigade-level example is of a much more complex range of considerations where in many (but not necessarily all) cases, part of the process for achieving good decisions may be the interplay of different human perceptions through several minds in discussion. AI may enhance the raw speed of the OODA loop but too great a reliance on this technology may rob the process of necessary moral complexity. The aim must be to speed up the OODA loop but to be discriminating about when absolute speed is a necessity (as with CIWS) and where only greater speed relative to an opponent is the necessary aim.

DISCRIMINATION AND PROPORTIONALITY

If we examine the problem of targeting and collateral damage, AI may have valid applications here too. The pilot of a drone can be located on the other side of the world, choosing whether or not to fire the drone's weapons. Imagine that a target has been identified and the drone operator must choose whether to fire on that target. There is a high risk of collateral damage and perhaps civilians moving in and out of the area. Our human is conscious of the Rules of Engagement (including international agreements like the Geneva Conventions) and understands that elimination of the target is a military necessity. The operator may be justified in firing when they feel that there is a lull in civilian activity but let us imagine that the target is next to a busy street – how well judged is that feeling? There may be side alleys and intersections to keep an eye on, as well as the fact that the platform from which the missile will be launched is moving, changing the travel time of the missile. Any opportunities to minimise civilian casualties will be fleeting and a human may not be best equipped to seize those opportunities.

AI could track the variables that are occurring, as well as track the location of the drone itself and compensate. When the conditions are met of minimal casualties, the AI could fire. A human would be firmly on the loop still and would essentially hand the moment of decision making over to the AI at some point so that fundamentally the AI has made the decision to take a human life. However, this may not be as problematic as it first seems. Without the AI, the decision to take that life had already been made and a human would pull the trigger which would arguably put other human life at greater risk. With the AI, the overall end state is the same, the target has still been killed, but there is an important nuance; the AI was given control, not so it could kill more effectively but so that civilian life could be better protected. The intention in deploying the AI was not fundamentally to kill (that had already been decided) but to minimise collateral damage. The use of AI in this instance seeks to contain the impact of wars, entirely in line with the principles of Just War Theory.

In the example, there is still a human on the loop – but what if this were not the case? Landmines, for instance, are indiscriminate, and used to block areas from enemy forces. However, a mine will detonate for a civilian just as much as for an enemy soldier, and their persistence after wars have ended has led to ongoing tragedies. Banned under the Ottowa Convention, several nations nevertheless continue to use them as a relatively inexpensive yet undeniably effective way to secure territory. AI offers the potential of making landmines more able to discriminate and, therefore, more compatible with the principles of *jus in bello*. A mine that could decide when to detonate could be far more humane, targeting combatants as opposed to children or other civilians. This would limit the excesses of war and still provide many of the benefits of landmines without many of their horrific disadvantages. Again, the intent matters; the landmines detonate in line with the principle of discrimination. The decision to kill has still ultimately been made by a human, the AI is just carrying it out in as proportionate and discriminating a way as possible.

It should be noted that a potential counter to this argument is one of unintended consequences. For example, it may encourage an increase in the use of child soldiers if they prove to be the ones who can get through the minefields. But it can also be argued that it is not a risk that 'our side' owns. Attempts to limit the use of landmines have been only partially successful. If they cannot be removed entirely, then there is a choice of leaving landmines as they are or attempting to make them more discriminating. That may be abused but the alternative, of not trying at all, seems worse. Making every effort to limit the excesses of war remains an excellent guiding principle, and AI represents a way of achieving that.

AI AND WEAPONS OF MASS DESTRUCTION

International humanitarian law has long sought to restrict the indiscriminate and disproportionate use of weapons (Crawford & Pert, 2024). Weapons of mass destruction

(WMDs) are, by their nature, indiscriminate and potentially massively more disproportionate. It is not surprising, then, that WMDs have seen even greater efforts to curb their development and use than mines (NATO, 2023). Given its widespread application, combined with our global reliance on networked computing and communication infrastructure, AI itself has the potential to be developed and deployed as a WMD (McEvoy, 2022). Paradoxically, its capacity to discriminate targets and impose parameters on its proportionality may counter-balance that potential but only to an extent: the risk is a political one rather than intrinsic to AI. The potential within AI to reinforce the principles of Just War should not obscure its simultaneous potential to be deployed in ways which breach those principles and enter the realm of WMDs.

Some jurisdictions are moving towards tackling the risks in AI through legislation. In late 2023, the European Union approved the EU AI Act, the first legislation to regulate the use of AI within the EU (Zhong et al., 2024). This regulation bans from use within the EU any AI that is deemed to be of unacceptable risk, and imposes strict regulation on AI that is deemed to be of high risk, but explicitly excludes military applications of AI from its scope. It is nonetheless clear that there are risks entailed by the use of AI in military applications comparable to those posed by WMDs. It is equally clear that the scope of AI extends well beyond the borders of the EU. We will be reliant – as with established WMDs – on brokering and enforcing international agreements and treaties.

Here arises another interesting paradox in that, in terms of negotiating treaties and agreements, adopting a national or monocultural position in isolation from the geopolitics of both allies and potential adversaries is unlikely to be adequate. As we have argued, a culturally specific ethical framework must be incorporated. The AI design process and the political processes for managing AI in military contexts demand different and complementary approaches to international relations.

OTHER BENEFITS AND RISKS

AI could also bring military advantages in electronic warfare while serving the Just War principles of discrimination and proportionality, for example in automatically jamming enemy communications far more efficiently. Fixed Frequency Hopping is a standard defence against electronic jamming and sees radios moving across multiple frequencies every second. With human-led jamming, the only effective counter is to jam every frequency available, which is power-intensive (limiting the effectiveness of the disruption) and can disrupt frequencies used by emergency services (potentially violating the principles of discrimination and proportionality), or to jam only specific frequencies which leads to incomplete jamming. An AI system could target specific frequencies far more effectively while identifying which frequencies would be best used by friendly forces against enemy jamming attempts.

But the Terminator in the room must also be addressed; the issue of so called ‘killer robots’. Let us suppose that we create an autonomous machine to replace tanks or personnel on the ground and give it the ability to decide to kill or not. Again, intent is significant here. If the intent is to kill as many people as possible then it is likely that an AI would be very good at this. The US military conducted a study after

World War Two into what it felt was a gap in ratios of fire, with the historian and US Army Officer SLA Marshall estimating that only 15 to 20 percent of riflemen fire their weapons in battle (Engen, 2011, p. 39), positing a number of reasons for this. While his estimates and reasons are now subject to some doubt, it is clear that many personnel did not fire their weapons or aimed to miss. AI may not have this compunction and perhaps one of the most unrealistic aspects of the Terminator films is that the machine misses so often. However, under the Just War principle of military necessity (May, 2007, p. 190), soldiers are not given orders to kill as many as possible; acts of violence are only carried out so far as they are necessary to secure an objective. Adopting the same ethic, our robots would not be ordered to kill as many as possible; it has no inherent benefit. Our AI would be given a task to achieve and it has the means to decide to use lethal force if it is necessary to achieve that end.

AI's ability to aim and fire at a target and hit it more often than not could also save lives. There would be fewer rounds firing as a result and, of those rounds that are fired, fewer rounds missing a target. That would mean less collateral damage. If greater accuracy limits collateral damage, then a machine that is more accurate with its weapons is an act of humanity.

Another argument is that machines do not experience some of humanity's more admirable qualities: empathy, compassion, love and sympathy, to name but a few. They will be heartless in executing their commands. However, humanity is a package deal. While humans are capable of great deeds, they are also capable of devastating lows. Humans can also be angry, desperate and scared, leading to illegal killings. This was exemplified by Alexander Blackman who was initially found guilty of murder for the killing of an Afghan insurgent in Helmand province in 2011. Blackman had been under considerable stress due to the situation on the ground at the time and the preceding activities that had made up his operational tour, and so his charge was later reduced to manslaughter on the grounds of diminished responsibility (Regina v Blackman, 2017). The stressors that he endured explain a great deal of why he committed manslaughter but would a machine be as impacted by those stressors or would they have no effect on it? Just as an AI may not feel sympathy, it will also not feel anger. There may be situations where cold rationality may be a better prospect. If the intent is to avoid unlawful killings and make the principles of discrimination and proportionality absolute, and the AI is used in accordance with military necessity, then humane outcomes may follow from using the non-human.

THE ETHICS OF AUTONOMY

Despite this potential for AI to correct for human error and human weakness, mitigating the randomness and chaos of the battlefield, in Just War terms there will remain justifiable concerns and questions over how much agency, decision making or control to hand over to AI. It may help to start with some thoughts about autonomy as a human characteristic since, if we are working with an inadequate account of humanity, we may end up ceding the most distinctive aspects of humanity to machines.

Indeed, two key aspects of being human which modernity has bequeathed to our self-perceptions – autonomy and rationality – turn out, on inspection, to be only part of the story. This is not the place for a fully developed anthropology but dependency is in many ways a greater truth about human existence than autonomy (MacIntyre, 1999, *passim*). By describing some AI weapons systems as autonomous, it may be that the chosen vocabulary itself serves to obscure the dependence of the machine on human input, both in design and in deployment.

Then again, while being human includes a capacity for rationality, absence of rationality or its qualification by other factors does not render one non-human. Nor is rationality the supreme human characteristic. As noted above, feelings and emotions – anger, empathy, love and sadness – are inseparable from being human. Eradicating human characteristics such as anger and fear through deployment of AI might be justified on Just War terms but human qualities such as empathy and mercy also need to be factored in. It is one of the weaknesses of classical utilitarianism that decisions may be presented as fair according to some rational calculus but do not always strike those affected as being fair at all. And this is surely because concepts such as fairness, including that of justice, are developed not from a rational assessment of a situation but from something much deeper – that is, the kind of moral person we perceive ourselves to be. Yet again this points to the difficulty of deploying AI in military contexts without building in the ethical traditions of the human user's culture. In the end, only having the human deeply involved in the design and deployment of an AI can bring to the equation human qualities such as mercy which are integral to practices of justice, including in war.

Concepts like living well and being good are not exhausted by arriving at the most efficacious answer to a problem. Thus, it is possible to speak, for example, of a 'good doctor' as a set of practices that run well beyond always getting the diagnosis correct (MacIntyre, 1982, p. 187). Just War Theory encapsulates some of the principles for being a good soldier or war leader. The notion of war crimes is the reciprocal understanding that not all military personnel are 'good' and that breaches of good practice should be punished. Although there is a great deal of international jurisprudence about the interpretation of the so called Nuremburg Defence (i.e. the plea that one was only obeying orders), it remains that while this defence has sometimes mitigated punishments it has not been especially successful in arguing that defendants were not responsible for outcomes. The Rome Statute of the International Criminal Court explicitly rules out the Nuremburg Defence in cases of genocide or crimes against humanity and in other cases which do not meet strict criteria determined case by case. Justice is not simply a matter for institutions to determine but a virtue for persons to practise (Porter, 2016, p. 227).

The weakness of the Nuremburg Defence suggests that if delegating human autonomy upwards to one's superior does not absolve one from moral responsibility for one's actions, so delegating autonomy to a machine running AI applications is likewise no adequate moral defence for what may follow. AI may, as we have shown, aid human decision making and action by enabling it to be faster, to encompass more data and to calculate more precisely. But moral responsibility must always be traceable back to human agency or else it would be impossible to practice *jus in bello*. The Just War tradition is for the guidance of human beings who seek to act well and pursue the good; it does not substitute for personal moral responsibility. The applications of AI in military contexts must therefore ensure the traceability of commands to human agency – enhanced by AI but not replaced by it – and this should be embodied in international law.

CONCLUSION

While the military traditions of the end user are largely irrelevant to the design of conventional weapons, in the case of AI-driven weaponry, there must be a partner in the process to speak up for the principles of *jus in bello* since market forces themselves, negligent of internal goods and reticent about cultural specificities, will not and cannot do so. Could such a process be enshrined in international law? This has been done, to a limited degree and with limited success, in the case of WMDs and other conventional weapons. There should be no reason why similar agreements cannot be forged in the case of AI. These are challenges for politics and diplomacy which are not unique to AI applications.

Many discussions about the use of AI in future scenarios focus on potentially deleterious consequences and unanswered questions. But there are many areas in which the application of AI suggests potentially hopeful and morally positive outcomes. Even in more offensive applications, limiting or reducing the indiscriminate or disproportionate use of weapons is an ethically significant potential application of AI.

Indeed, we conclude that the use of AI can be 'good' or ethical in both defensive and some offensive contexts, according with and sometimes enhancing adherence to the principles of *jus in bello*. It is arguable that it could sometimes be immoral not to use it. In some scenarios there is a moral argument to hand over decision making and acting to AI in situations where a human cannot be as effective. When we do that, AI becomes not just another weapon. Provided the design, development and deployment of AI is underpinned by the values of Just War Theory, it may be ethically acceptable and indeed potentially ethically imperative. But there is a lot of political, conceptual and practical work to be done to achieve those conditions.

REFERENCES

Bagaric, M., Svilar, J., Bull, M., Hunter, D., & Stobbs, N. (2022). The solution to the pervasive bias and discrimination in the criminal justice system: transparent and fair artificial intelligence. *American Criminal Law Review*, *59*, 95.

Crawford, E., & Pert, A. (2024). *International humanitarian law* (3rd ed.). Cambridge University Press.

Dalton-Brown, S. (2020). The ethics of medical AI and the physician-patient relationship. *Cambridge Quarterly of Healthcare Ethics*, *29*(1), 115–121.

Engen, R. (2011). SLA Marshall and the ratio of fire: History, interpretation, and the Canadian Experience. *Canadian Military History*, *20*(4), 39–48.

Flores, L., Kim, S., & Young, S. D. (2024). Addressing bias in artificial intelligence for public health surveillance. *Journal of Medical Ethics*, *50*(3), 190–194.

Gichoya, J. W., Thomas, K., Celi, L. A., Safdar, N., Banerjee, I., Banja, J. D., … Purkayastha, S. (2023). AI pitfalls and what not to do: mitigating bias in AI. *The British Journal of Radiology*, *96*(1150), 20230023.

Gundersen, T., & Bærøe, K. (2022). The future ethics of artificial intelligence in medicine: Making sense of collaborative models. *Science and Engineering Ethics*, *28*(2), 17.

Hayek, F. A. (1944). *The Road to Serfdom*. London: Routledge & Kegan Paul.

Krepon, M. (2021). *Winning and Losing the Nuclear Peace: The Rise, Demise, and Revival of Arms Control*. Stanford, CA: Stanford University Press.

Langan, J. (1984). The Elements of St. Augustine's Just War Theory. *The Journal of Religious Ethics*, *12*(1), 19–30.

MacIntyre, A. (1982). *After Virtue: A Study in Moral Theory* (2nd ed.). London: Duckworth.
MacIntyre, A. (1999). *Dependent Rational Animals: Why Human Beings need the Virtues*. London: Duckworth.
May, L. (2007). *War Crimes and Just War*. Cambridge: Cambridge University Press.
McEvoy, J. (2022). Artificial Intelligence: The Ultimate Deterrence. *Artificial Intelligence: The Ultimate Deterrence* » Wavell Room. Accessed: 26 March 2024.
Miller, S. (2022). AI and Criminal Justice. In M. Boylan & W. Teays (Eds.), *Ethics in the AI, Technology, and Information Age* (pp. 207–231). Maryland: Rowman and Littlefield.
Ministry of Defence. (2022). *Defence Artificial Intelligence Strategy*. Accessed: 22 April 2024.
NATO. (2023). Arms control, disarmament and non-proliferation in NATO. Retrieved from https://www.nato.int/cps/en/natohq/topics_48895.htm
Peters, U., & Carman, M. (2024). Cultural bias in explainable AI research: a systematic analysis. *Journal of Artificial Intelligence Research, 79*. https://doi.org/10.1613/jair.1.14888
Porter, J. (2016). *Justice as a Virtue: A Thomist Perspective*. Grand Rapids, MI: William B. Eerdmans.
Ramsey, P. (1961). *War and Christian Conscience*. Durham, NC: Duke University Press.
Reed, C. (2004). *Just War?*. London: SPCK.
Regina v Blackman, [2017] EWCA Crim 190, In the Court Martial Appeal Court On A Reference By The Criminal Cases Review Commission On Appeal From A Court Martial At Bulford, R -v- Alexander Wayne Blackman - judgment. Retrieved from https://www.judiciary.uk/judgments/r-v-blackman/
Richards, C. (2020). Boyd's OODA Loop. *Necesse*, *5*(1), 142–165.
Shahbazi, N., Lin, Y., Asudeh, A., & Jagadish, H. V. (2023). Representation bias in data: a survey on identification and resolution techniques. *ACM Computing Surveys*, *55*(13s), 1–39.
Temple, W. (1941). *Citizen and Churchman*. London: Eyre & Spottiswoode.
Woodcock, T. K. (2024). Human/machine(-learning) interactions, human agency and the International Humanitarian Law proportionality standard. *Global Society*, *38*(1), 100–121. https://doi.org/10.1080/13600826.2023.2267592
Zhong, H., O'Neill, E., & Hoffmann, J. A. (2024). Regulating AI: applying insights from behavioural economics and psychology to the application of Article 5 of the EU AI Act. *Proceedings of the AAAI Conference on Artificial Intelligence*, *38*(18), 20001–20009. https://doi.org/10.1609/aaai.v38i18.29977

4 Lethal Autonomous Weapons and the Technocratic Paradigm

Matthew J. Gaudet

This is a book on the implications of the family of technologies known collectively as artificial intelligence (AI) and the tradition of thought known as the just war theory (JWT). Both of these terms—artificial intelligence and just war theory—are wide ranging. As others in this volume have skilfully demonstrated, there is a wide array of ways in which AI can be applied to warfare, from intelligence gathering to code-breaking to social destabilizing disinformation campaigns. While each application of AI to warfare deserves its own ethical analysis, for this chapter, I will limit my attention to the subject of lethal autonomous weapons systems (LAWS).

Similarly, the just war is not a monolithic set of rules for ethical war-making. Rather, it is a tradition of thought, with numerous branches and rival perspectives, drawn from Catholic and Christian theology and various secular philosophers, on what constitutes a just war. The tradition is generally dated back as far as St. Augustine of Hippo in the 4th Century, and traces through the Roman-Christian-European thought to the 20th Century. Its most significant advances occur with the 13th Century Christian scholastics (e.g. Thomas Aquinas) and then again in the hands of the 15th and 16th Centuries jurists (e.g. Francisco Suarez, Francisco de Vittoria, Hugo Grotius). The form of the just war tradition that is most familiar today emerged in the late 20th Century (Gaudet & O'Neill, 2011). Philosopher Michael Walzer's landmark 1977 text, *Just and Unjust Wars* is widely recognized as the summative text that brought many of the concepts of the just war—such as the *jus ad bellum* (the justice of [declaring] war itself) and the *jus in bello* (just means used in a conflict)—into the popular discourse (Walzer, 1992). Soon after, it was James Childress, applying a *prima facie* deontological argument in 1980 that first named the now familiar structure of 6 or 7 *jus ad bellum* criteria and 2 or 3 *jus in bello* criteria for determining if a war is just (Childress, 1980), but it was the United States Conference of Catholic Bishops' 1983 pastoral letter *The Challenge of Peace* that would popularize these criteria (United States Catholic Conference of Bishops, 1983). Once again, for my contribution to this volume, I will not attempt to draw upon the entirety of this tradition, but instead operate within the confines of the *Roman Catholic* tradition of thought on just war thinking.

So, within this book on Artificial Intelligence and Just War, this particular chapter will examine how Catholic just war applies to lethal autonomous weapons systems.

DOI: 10.1201/9781003440253-4

However, one further specification needs to be made. The Catholic tradition did not end with the 1983 American Bishops letter and in fact, the current Catholic pope, Francis, has developed a robust technological ethic that advances Catholic ethical and social thought (including attention to AI) and must, therefore, be brought into conversation with just war thinking to properly address the moral questions of LAWS. In summary then, my argument in this chapter is that ethical analysis of lethal autonomous weapons systems using the Catholic just war tradition must first account for Pope Francis's technology ethic and in particular, his identification of a 'technocratic paradigm' that is operating and restraining human relationships in contemporary society.

THE CATHOLIC JUST WAR TRADITION

The *Challenge of Peace* began with the understanding that nation-states have a moral responsibility to both do the good (including making the world more just) and to do no harm (non-maleficence). The problem is that even though war is inherently maleficent, sometimes our duty to justice requires that we have to (regrettably) conduct war. It is in these cases—where justice and non-maleficence are held in tension—that we need a guide to help us decide which commitment is more pressing. The criteria named in the *Challenge of Peace* were a framework for knowing when the needs of justice outweighed the commitment to non-maleficence.

In the Catholic account, war can only be justified when there is a **just cause**, an acute situation of injustice that demands a response. All other means of rectifying that injustice must be exhausted and war must only come as a **last resort**. War can only be declared by a **legitimate authority**. Under the modern nation state system, we do not want the city of El Paso declaring war on Mexico. Or today, when companies are as wealthy as nation states, we do not want Google declaring war on Canada. If there be injustices on both sides of the conflict, then war can only be justified in defence of the side suffering **comparatively greater injustice**. War must always be aimed a **right intention**, a desired future state that rectifies the injustice that gave us just cause but also minimizes harm on both sides of the conflict. Relatedly, the decision to go to war must be **proportionate**, balancing the injustice war aims to overcome against the inevitable harms that war itself will do. Finally, war should not be engaged unless there is a **reasonable hope** of achieving the right intention with proportionate harms. These are the 7 *jus ad bellum* (justice of the war) criteria that determine whether war itself is just.

The *jus in bello* (justice in war) criteria determine the proper means by which war should be fought. Here again, we must conduct war with **right intention**, and only use those means which will further that intention. Means of war that aim at personal gain, revenge, or blatant destruction are unjust. Similarly, the means of war must be **discriminate**. Civilians ought to never be directly targeted and means of war that do not discriminate are unjust. Finally, the means of war must be **proportional** to their immediate gains. We need not use intercontinental ballistic missiles in response to a small group of rebel terrorists. But also, when proportionality and discrimination are considered together, then in cases where harm to civilians is unavoidable, such harms must be reduced as much as possible.

The problem is that many who use the just war criteria today do so completely detached from the longer religious and philosophical tradition that both preceded it and continues to develop. They take the criteria themselves as a fixed set of 'rules for war.' When the criteria are separated from the 'thick' tradition, such rules are rendered so 'thin' to be almost meaningless.

American President Barack Obama, in his Nobel Peace Prize address, referred to the tradition in such 'thin' terms: '[W]ar is justified only when certain conditions were met: if it is waged as a last resort or in self-defense; if the force used is proportional; and if, whenever possible, civilians are spared from violence.' (Obama, 2009) Obama offered no presumption against violence; no tragedy in the decision to go to war; and no moral complication of weighing justice against non-maleficence. Just be sure you 'check the boxes' before sending troops in. And there was certainly no deliberation about what the meaning of those checks might look like in real life. Take just that final clause: civilians being spared (discrimination) is a just goal, but determining when an individual is a combatant and when they are a civilian is more complicated in practice. This is why Walzer spent four whole chapters of *Just and Unjust Wars* unpacking the nuances of this question, thickening the principle of discrimination by examining the nuances of cases such as firing at soldiers in unsoldierly positions (e.g. taking a bath or separated from their platoon) or the use of blockades that can affect civilians and soldiers alike. Walzer recognized that thin principles are insufficient. We need thick accounts behind the thin principles to give them actual moral force. Walzer's thickened his account through a casuistic methodology that richly described historical cases in order to extract communitarian norms. Catholic just war, by contrast, thickens the just war principles by rooting it within the Catholic tradition and its fundamental commitments to human dignity, love of neighbour, care for the most vulnerable, and promoting the common good.

For its part, the *Challenge of Peace* is rather thin in its commitments (the document only reserves three paragraphs for describing discrimination, for example), but the letter was never intended to stand on its own. The opening words of the document quote the Second Vatican Council document *Gaudium et Spes* and thus link the *Challenge of Peace* to the broader tradition of Catholic social thought and the document as a whole is littered with references to the full tradition of Catholic moral theology. Thus, while the direct treatment of a principle like discrimination was only a few paragraphs, it is built upon a commitment to the fundamental need to protect human life and human dignity that was nearly 2000 years in the making. The point, for the purposes of this chapter, is that Catholic just war thinking is inseparable from a much wider tradition of Catholic thought and teaching. Thus, for us to bring the just war tradition to bear in assessing the legitimacy of Lethal Autonomous Weapons Systems, we must do so against a thick tradition that has developed over centuries and continues to develop today. For our purposes here, I will not attempt to rehash the entirety of Catholicism's moral commitments, but the recognition that the just war emerged from a larger Catholic tradition necessarily brings the present chapter on LAWS into conversation with the current Roman Catholic Pope, Francis and his response to the state of technological innovation in the 21st Century, including the advent of artificial intelligence.

POPE FRANCIS AND THE TECHNOCRATIC PARADIGM

Pope Francis embraces the ability to conceive and create, especially in the form of technology, as an expression of our humanity: 'The Church never ceases to encourage the progress of science and technology placed at the service of the dignity of the person, for an "integral and integrating" human development' (Francis, 2023). Francis is not anti-technology, but he is critical of the *relationship* we have with technology in the 21st century and what that relationship has done to human flourishing in our contemporary world.

Francis ascended to the papacy in 2013. In his twelve years in the role, he has issued four papal encyclicals—letters written with the highest authority of the office and often intended to speak to a pressing global topic. Many encyclicals, though part of the Catholic tradition, are not intended merely for the Catholic faithful, but to engage the world as a whole in conversation about a topic of global concern. Francis's predecessors had used encyclicals to address working conditions during the industrial revolution (*Rerum Novarum*), the economic conditions of the Great Depression (*Quadragesimo Anno*), and Cold War hostilities (*Pacem in Terris*). In this spirit, Francis aimed his second encyclical, *Laudato Si'* at the climate crisis. However, in the process of addressing this topic, Francis also began to articulate a particular account of the relationship we have with technology in the contemporary industrialized world.

Technology should be a tool we use to achieve other, intrinsically valuable, ends: better health, better communication, stronger human relationships, more time for leisure and enjoyment, etc. What Francis observed, however, was that in our contemporary world, technology no longer served such an instrumental purpose, but instead had become an end unto itself.

> Technology tends to absorb everything into its ironclad logic, and those who are surrounded with technology know full well that it moves forward in the final analysis neither for profit nor for the well-being of the human race, that in the most radical sense of the term power is its motive – a lordship over all. As a result, man seizes hold of the naked elements of both nature and human nature. Our capacity to make decisions, a more genuine freedom and the space for each one's alternative creativity are diminished.
>
> (Francis, 2015, para. 108)

We desire new technology, not because it is useful, but because it is new. We want the iPhone with the latest camera, even though the marginal difference from the phone in our pocket is hardly perceptible by the human eye. We replace our washing machine, refrigerator, or toothbrush with one that is 'smart' and internet connected, even if we never use these 'features' in real life. And, in the past two years, every company wants to be able to claim that their tech utilizes 'artificial intelligence' even if biases, AI hallucinations, or a lack of overall capability actually makes the technology worse than the systems being displaced.

We seek 'progress' for progress's sake, 'disruption' for disruption's sake, and 'technology' for technology's sake, without asking any real questions about the benefits of the new technology, let alone the costs an externalities:

> There is a tendency to believe that every increase in power means 'an increase of 'progress' itself', an advance in security, usefulness, welfare and vigour; …an assimilation of new values into the stream of culture, as if reality, goodness and truth automatically flow from technological and economic power as such. The fact is that 'contemporary man has not been trained to use power well', because our immense technological development has not been accompanied by a development in human responsibility, values and conscience. Each age tends to have only a meagre awareness of its own limitations. It is possible that we do not grasp the gravity of the challenges now before us.
>
> (Francis, 2015, para. 105)

In a similar way, we seek profit for profit's sake, increasing the coffers of corporations without any commitment that such profits will be reinvested in technological progress directed toward some good end.

As technology, progress, disruption and profit turn from an instrumental good to an intrinsic good, those things that ought to be valued intrinsically—human life and the environment in which we live—are simultaneously reduced to a mere resource to be consumed. 'Indeed, we seem to be losing the value and profound meaning of one of the fundamental concepts of the West: that of the human person.' (Francis, 2024b) Consider the implications of the commonly used terms 'human resources' and 'natural resources' in business practice. The fundamental human dignity that should be most precious to us has become a mere utility, to be used while useful, then discarded. Relatedly, 'progress' is increasing measured in individual terms, rather than recognizing and promoting the existence of a common good that is shared across communities and cannot be possessed by individuals.

For Francis, none of these individual social shifts are all part of a single 'technocratic paradigm' that is deeply embedded in the structures and culture of the modern digital world (AI Research Group for the Centre for Digital Culture, 2024). This paradigm sets our values and shapes our decisions. It determines everything from how engineers design, how marketers market, and how users use technology all the way to the discarding and disposal of that same technology (an important theme in a document on the climate crisis). In *Laudato Si'*, Francis names the technocratic paradigm as one of the major hurdles necessary to overcome to get to the collective action necessary to respond to climate change but his third encyclical, *Fratelli Tutti*, broadens the impacts of the technocratic paradigm.

Francis addressed the breakdown in social cohesion that we are witnessing from the international down to the interpersonal level in the 21st century. Here again, he points to the technocratic paradigm as an underlying barrier to solving the problems:

> Some parts of our human family, it appears, can be readily sacrificed for the sake of others considered worthy of a carefree existence. Ultimately, persons are no longer seen as a paramount value to be cared for and respected, especially when they are poor and disabled, 'not yet useful' – like the unborn, or 'no longer needed' – like the elderly. We have grown indifferent to all kinds of wastefulness, starting with the waste of food, which is deplorable in the extreme.
>
> (Francis, 2020b, para. 18)

Here he extends the analysis of technology to argue that we have a practice of placing ever more extensive technology between ourselves and those around us and that this practice has had deleterious effects on our individual social bonds and our broader social fabric. We see this implicitly in forms of communication: text is a limited medium compared to voice and videoconferencing and teleconferencing is inferior to in person discussion.

> They lack the physical gestures, facial expressions, moments of silence, body language and even the smells, the trembling of hands, the blushes and perspiration that speak to us and are a part of human communication. Digital relationships, which do not demand the slow and gradual cultivation of friendships, stable interaction or the building of a consensus that matures over time, have the appearance of sociability. Yet they do not really build community; instead, they tend to disguise and expand the very individualism that finds expression in xenophobia and in contempt for the vulnerable. Digital connectivity is not enough to build bridges. It is not capable of uniting humanity.
>
> (Francis, 2020b, para. 43)

By the same token, we have all come to see the problems of social media as a platform, from unrealistic portrayals of other's lives on social media lead to depression and low self-esteem to algorithmic sorting leading to deep polarization. Placing social media between us as the medium through which our relationships communicate harms rather than strengthens our interactions:

> Dialogue is often confused with something quite different: the feverish exchange of opinions on social networks, frequently based on media information that is not always reliable. These exchanges are merely parallel monologues. They may attract some attention by their sharp and aggressive tone. But monologues engage no one, and their content is frequently self-serving and contradictory.
>
> (Francis, 2020b, para. 200)

These are merely a few of the ways in which, in *Fratelli Tutti*, Francis raises concerns around the role technology plays in contemporary culture and the universal effect that is has in cutting us off from our connections to nature and our connections to each other. In particular, he is concerned that we are not 'putting science and technology at the service of peoples' (Francis, 2016). This is just as true of Artificial Intelligence as it is of conventional technology.

Since 2020, Francis (along with the rest of the world) has turned his focus toward artificial intelligence and its effects on society. But his overall concerns about technology serving as an end in itself are only exacerbated by the AI revolution:

> 'Intelligent' machines may perform the tasks assigned to them with ever greater efficiency, but the purpose and the meaning of their operations will continue to be determined or enabled by human beings possessed of their own universe of values.
>
> (Francis, 2024a)

For Francis, those values begin with the dignity of the human person and the strengthening of the social bonds of humanity:

> Artificial intelligence ought to serve our best human potential and our highest aspirations, not compete with them. …Technological developments that do not lead to an improvement in the quality of life of all humanity, but on the contrary aggravate inequalities and conflicts, can never count as true progress.
>
> (Francis, 2024a)

While the potential positive impacts of artificial intelligence are extensive and ought to be pursued. However, the introduction of Artificial Intelligence to society without due care and concern for the human outcomes, especially for the most vulnerable in society, has the potential to exacerbate our greatest social flaws. In the coming section, I will show how this same argument extends to the use of AI in Autonomous Weapons Systems and how a thick understanding of both the technocratic paradigm and the just war tradition are necessary to providing such care and concern.

LETHAL AUTONOMOUS WEAPONS AND JUST WAR

When he wrote *Fratelli Tutti* in 2020, Francis was already so concerned that the *capacity* of modern weaponry had become so destructive that hope in the limits of a just war may well be rendered impossible.

> [T]he enormous and growing possibilities offered by new technologies, have granted war an uncontrollable destructive power over great numbers of innocent civilians. The truth is that never has humanity had such power over itself, yet nothing ensures that it will be used wisely. We can no longer think of war as a solution, because its risks will probably always be greater than its supposed benefits. In view of this, it is very difficult nowadays to invoke the rational criteria elaborated in earlier centuries to speak of the possibility of a 'just war.'
>
> (Francis, 2020b, para. 258)

At that time, however, his thought on military ethics was still focused on the move to drone warfare that had occurred over the past two decades and significantly removed those who pull the trigger from direct contact with the enemy. Since that time, the public at large has been witness to several forms of AI. Generative AI systems, such as ChatGPT, have captured public attention recently, but in 2021, a United Nations Security Council report strongly suggested that during a March 2020 attack on the militia forces of local warlord Khalifa Haftar, Libyan government forces used a swarm of Turkish-made Kargu-2 drones, operating without any human operator or input (United Nations Security Council, 2021). The Libyan government has denied the report, but if confirmed as true, it would be the first attack in history in which was humans were killed by a completely autonomous AI system.

In 2024, Pope Francis dedicated his World Day of Peace Message to the contributions of Artificial Intelligence for (and against) the task of peacemaking. For Francis, human dignity remains the central context for just war thinking:

> The inherent dignity of each human being and the fraternity that binds us together as members of the one human family must undergird the development of new technologies and serve as indisputable criteria for evaluating them before they are employed, so that digital progress can occur with due respect for justice and contribute to the cause of peace.
>
> (Francis, 2024a)

He is also cautious that new technology necessarily equates to human progress and, in particular for warfare, greater peace:

> We cannot presume a priori that its development will make a beneficial contribution to the future of humanity and to peace among peoples. That positive outcome will only be achieved if we show ourselves capable of acting responsibly and respect ...human values.
>
> (Francis, 2024a)

Most importantly, however Francis recognizes that our consideration of Autonomous Weapons (and war in the 21st Century in general), has to recognize the technocratic paradigm that is imbedded in contemporary society and shapes our understanding of technology, war, and especially the technology of warfighting.

War has always been an inherently 'human' activity. It is humans that go to war, for very human reasons. It is humans that decide when and how to shoot, and humans that live or die as a result. Those who advocate *for* autonomous weapons argue that these human aspects are what make war so tragic. Removing human soldiers means reducing human casualties. But even more than that, removing human decision making reduces the chance of decisions made because of human weaknesses. LAWS will never act out of revenge or anger. Presumably they should never be programmed to rape or torture. LAWS will be programmed with an objective (a right intention) and once that objective is met, they will cease the offensive. By this logic, fighting wars with LAWS would be *more humane* than conventional means. At the same time, LAWS will not have some of the weaknesses humans bring to the battlefield. LAWS will not hesitate to act out of fear, panic, or indecisiveness. LAWS will not be limited to the five human senses but can collect data from extensive sensors and process it at a speed orders of magnitude faster than humans (AI Research Group for the Centre for Digital Culture, 2024, pp. 202-3).

However, turning decision-making over to machines means that we also remove from the battlefield other inherently human traits. For Francis, this begins with the capacity for moral responsibility:

> Autonomous weapon systems can never be morally responsible subjects. The unique human capacity for moral judgment and ethical decision making is more than a complex collection of algorithms, and that capacity cannot be reduced to programming a machine, which as 'intelligent' as it may be, remains a machine. For th is reason, it is imperative to ensure adequate, meaningful and consistent human oversight of weapon systems.
>
> (Francis, 2024a)

There is a practical and legal question regarding responsibility: if a LAWS system (or any AI system) does something criminal, who bears responsibility? And if the system is found liable, what is the recourse?

Even beyond the legal questions, how does the presence of responsibility change how we fight wars? If we remove the human capacity for compassion, empathy, and justice from the battlefield, what will be the ramifications for the *jus in bello*? LAWS will never be capable of recognizing the reflection one's own human dignity in the presence of any enemy and it is this recognition that provides thick understanding of the principles of discrimination or proportionality. In the Catholic just war tradition, war is always tragic precisely because every human person—even our enemies—is imbued with a fundamental human dignity and a fundamental right to life. At times, war is the lesser tragedy and thus necessary, but it always must remain tragic and thus limited, proportional, and discriminate It is a vital question whether artificial intelligence systems are capable of recognizing tragedy. Given AI's foundation as algorithmic programming, recognition of abstract ideas such as tragedy, human dignity, or love would seem beyond AI's capacity. As such, even if the technology did prove capable of some algorithmic calculation of discrimination or proportionality, it would not be the same as the version of these principles that is thickly constituted in the human experience and human dignity.

For all of these reasons, Francis ultimately concludes, 'In light of the tragedy that is armed conflict, it is urgent to reconsider the development and use of devices like the so-called 'lethal autonomous weapons' and ultimately ban their use....No machine should ever choose to take the life of a human being.' (Francis, 2024b)

THE TECHNOCRATIC PARADIGM, *JUS AD BELLUM*, AND LAWS

But the just war analysis of LAWS do not conclude with the application of the *jus in bello* principles. There are important *jus ad bellum* concerns with LAWS as well. As I have noted, for almost a decade, the Pope has expressed concern over the ways in which technology tends to diminish rather than support human flourishing and human relationship. While it may seem an unexpected application, I would argue that this general caution about the role technology plays in our modern relationships applies, in its own way, to the relationship between military adversaries. The more technology we place between ourselves and our enemies, the easier it is to dehumanize them, make light of the tragic act of killing, turn warfighting into a cost-benefit analysis, and, to put it most generally, remove the human elements that help to restrain warfare.

This phenomenon is not new to warfighting, but modern technologies that remove soldiers from direct contact with enemy soldiers have exacerbated the problem:

> [W]e have certain ancestral fears that technological development has not succeeded in eliminating; indeed, those fears have been able to hide and spread behind new technologies. Today too, outside the ancient town walls lies the abyss, the territory of the unknown, the wilderness. Whatever comes from there cannot be trusted, for it is unknown, unfamiliar, not part of the village. It is the territory of the 'barbarian', from whom we must defend ourselves at all costs. As a result, new walls are erected for self-preservation, the outside world ceases to exist and leaves only 'my' world, to the point that others, no longer considered human beings possessed of an inalienable dignity, become only 'them'. Once more, we encounter 'the temptation to build a culture of walls, to raise walls, walls in the heart, walls on the land, in order to prevent this encounter with other cultures, with other people.
>
> (Francis, 2020b, para. 27)

Holding on to the human dignity of our enemies is part of what separates the Catholic just war from those philosophies (e.g. Machiavellianism, realpolitik) that see war as completely outside of the boundaries of moral limits and only measurable in terms of gains and losses. An overly thin reading of the just war tradition unwittingly supports these philosophies because it reduces the 'rules of war' to a mere checklist that, once checked, returns decision making to the cost-benefit framing.

> By contrast, Catholic just war upholds a fundamental belief that war is tragic, and thus it must be avoided unless the very last resort; it must be prompted by a just cause; and it must be pursued with right intention and comparative justice. The question for us is how LAWS will challenge the practice of these commitments. In a 2020 address to the United Nations, Pope Francis offered his first direct critique of LAWS:We need to break with the present climate of distrust. At present, we are witnessing an erosion of multilateralism, which is all the more serious in light of the development of new forms of military technology, such as lethal autonomous weapons systems which irreversibly alter the nature of warfare, detaching it further from human agency. We need to dismantle the perverse logic that links personal and national security to the possession of weaponry. This logic serves only to increase the profits of the arms industry, while fostering a climate of distrust and fear between persons and peoples.
>
> (Francis, 2020a)

LAWS have emerged in the context of a society ruled by the 'perverse logic' of the technocratic paradigm that inverts our values and displaces the importance of human dignity, human flourishing, and community. In business, the technocratic paradigm orients us toward profit and 'progress' for their own sake, rather than as tools to be used to further our wellbeing. In the military sphere, however, it is the 'possession of weaponry' that has become an end in itself.

The technocratic paradigm shapes not just the expansion of military technology but the very decision to go to war. Consider that political scientists have argued that democracies do not tend to go to war with other democracies because those who bear the casualties of war—physical and psychological—the people themselves, hold the ultimate voting power in a democracy (Babst, 1964). The development of remote drones already put this theory to the test, and citizens were more likely to support a military operation if it was conducted by drone strike and thus did not risk (physical) casualties to one's own soldiers (Walsh & Schulzke, 2015). When LAWS do the fighting and warfighting no longer involves human casualties of any sort on our own side, what will become of *jus ad bellum* commitments to last resort, just cause, right intention, and comparative justice? Will LAWS create a frictionless slide toward war every time an enemy state missteps? Similarly, will the advent of LAWS systems exacerbate global inequalities in the world? The development of the most sophisticated weapons systems will allow the most wealthy and powerful states to engage war without human cost, while weaker states must make the choice between capitulation or human sacrifice for the sake of self-defense? LAWS are not merely a new type of weapon in a long history of development. The threshold crossed by fully automating warfighting decisions is one that could change war and society forever.

CONCLUSION

In the end, Pope Francis has been clear that he considers LAWS to be a grave violation of morality that ought not exist. However, if the pontiff's prohibition is not enough to stop these weapons from being developed (and presumably it will not be), Catholic just war and Catholic social thought (including Francis's development of the technocratic paradigm) have still more to add to a conversation regarding the ways in which these weapons can and should be limited. Just war has taught us that the deployment of the means of war must always be limited by the principles of discrimination and proportionality, thickly construed. The burden of these principles, has typically fallen on the soldier who pulls the trigger, and prior cases where the decision is automated (such as landmines) have been banned for their indiscrimination. LAWS promise to be more capable of discrimination than landmines ever were, but AI's capacity for this has not yet been proven. When it comes to proportionality, little is being discussed about including such a principle into the design of LAWS, but it needs to be.

But the concerns about LAWS extend beyond the *jus in bello*, back to the *jus ad bellum*. That is, as we continue to develop these and all systems of warfare, we must push back on the inclination to create new technology for technology's sake and weapons for the sake of being weaponized. If we concede to the promise of LAWS to making war more humane, we must remain aware that humans in the loop have always been a natural restraint on war. And, if we create AI systems that are designed to distance us from others (even our enemies), we must be very cautious that such means do not contribute to the type myopia, social isolation, and cultural insularity that are at the cultural root of many wars.

REFERENCES

AI Research Group for the Centre for Digital Culture. (2024). *Encountering Artificial Intelligence: Ethical and Anthropological Explorations.* (M. J. Gaudet, N. Herzfeld, P. Scherz, & J. J. Wales, Eds.) Eugene, Oregon: Pickwick.

Babst, D. V. (1964). Elective Governments—A Force For Peace. *The Wisconsin Sociologist, 3*(1), 9–14.

Childress, J. (1980). Just War Criteria. In T. Shannon, *War or Peace, The Search for Answers.* New York: Orbis.

Francis. (2015). *Laudato Si'*. Retrieved April 2024, from vatican.va: https://www.vatican.va/content/francesco/en/encyclicals/documents/papa-francesco_20150524_enciclica-laudato-si.html

Francis. (2016, November 5). *Address of His Holiness Pope Francis to Participants in the 3rd World Meeting of Popular Movements.* Retrieved April 2024, from vatican.va: https://www.vatican.va/content/francesco/en/speeches/2016/november/documents/papa-francesco_20161105_movimenti-popolari.html

Francis. (2020a, September 25). *Address to the United Nations.* Retrieved May 2024, from holyseemission.org; https://holyseemission.org/contents/news/5f6df2d493a95.php

Francis. (2020b). *Fratelli Tutti.* Retrieved from vatican.va: https://www.vatican.va/content/francesco/en/encyclicals/documents/papa-francesco_20201003_enciclica-fratelli-tutti.html

Francis. (2023, February 20). *Address of His Holiness Pope Francis to the Members of the Pontifical Academy for Life.* Retrieved May 2024, from vatican.va: https://www.vatican.va/content/francesco/en/speeches/2023/february/documents/20230220-pav.html

Francis. (2024a, January 1). *Message of His Holiness Pope Francis for the 57th World Day of Peace.* Retrieved from vatican.va: https://www.vatican.va/content/francesco/en/messages/peace/documents/20231208-messaggio-57giornatamondiale-pace2024.html

Francis. (2024b, June 14). *Address of His Holiness Pope Francis to the G7 Session on Artifucual Intelligence.* Retrieved from vatican.va: https://www.vatican.va/content/francesco/en/speeches/2024/june/documents/20240614-g7-intelligenza-artificiale.html

Gaudet, M., & O'Neill, S. W. (2011). Restoring Peace: Toward a Conversation Between the Just War and Reconciliation Traditons. *Journal for the Society of Christian Ethics, 31*(1), 37–55.

Obama, B. (2009, December 10). *Remarks by the President at the Acceptance of the Nobel Peace Prize*. Retrieved May 2024, from Obama White House Archives: https://obamawhitehouse.archives.gov/the-press-office/remarks-president-acceptance-nobel-peace-prize#

United Nations Security Council. (2021, March 8). *Letter from the Panel of Experts on Libya established pursuant to resolution 1973 (2011) addressed to the President of the Security Council*. Retrieved May 2024, from UN.org: https://documents.un.org/doc/un.oc/gen/n21/037/72/pdf/n2103772.pdf?token=9hHPtn17diKTt1k1m4&fe=true

United States Catholic Conference of Bishops. (1983). *The Challenge of Peace: God's Promise and Our Repsonse*. Washington: US Catholic Conference.

Walsh, J. I., & Schulzke, M. (2015, September). *The Ethics of Drone Strikes: Does Reducing the Cost Encourage War?* Retrieved May 2024, from U.S. Army War College: https://apps.dtic.mil/sti/pdfs/ADA621793.pdf

Walzer, M. (1992). *Just and Unjust Wars: A Moral Argument With Historical Illustrations* (2nd ed.). New York: Basic Books.

5 Above and Beyond

Drones, Artificial Intelligence and Just War Theory

James Garcia

This chapter will explore the ethical questions being raised regarding two of the most recent challenges to Just War Theory: drones and AI-enabled autonomous weapons. The continuing operations against global terrorism and the wars in Ukraine and Gaza provide relevant focus points to consider specific Just War *jus in bello* concepts such as non-combatant discrimination and proportionality. From the semi-autonomous Phalanx system used to protect U.S. Navy ships from incoming missiles to Israel's Iron Dome antimissile defence and Gospel target-generating systems, technology that uses machine learning and advanced computing is forcing reassessments of *jus in bello* beyond traditional concepts. The continuing use of drones in non-battlefield environments and lethal autonomous weapon systems in highly populated urban areas present new challenges to Just War thinking. This chapter will support the argument that traditional Just War concepts can still provide useful approaches to exploring the ethical questions raised by these newest additions to combatant arsenals. Their usefulness resides in that nexus of ideas and actions that join historical theories to international treaty and customary laws, and result in rational and functional laws of armed conflict and military rules of engagement.

As this chapter is being written, the war in Ukraine has witnessed a shift in air warfare from piloted jet fighters and bombers to the increased use of drones – large and small – as the primary airborne weapon of choice for both sides of the conflict. In its various forms, drone warfare is here to stay. The military use of uncrewed aerial vehicles (UAVs) or remotely-piloted aircraft (RPAs) has quickly moved from benign reconnaissance flights to lethal attack missions. What are the ever-evolving rules of engagement for this still relatively new form of combat? Are these rules consistent with the Just War tradition that has informed Western military thought since at least the mid-nineteenth century?

CONTEXT

Some writers argue for a new way of thinking about remotely piloted aircraft that builds upon Just War principles such as discrimination and proportionality. Yet others argue for moving beyond concepts of waging a just war *(jus in bello)* and instead

DOI: 10.1201/9781003440253-5

developing guidelines for achieving justice when using force-short-of-war *(jus ad vim)* or even perpetual force *(vim perpetua)*. While the continuing use of drones to target terrorist forces in non-battlefield environments has tested the limits of the Just War tradition, further discussion and reflection are needed to align reality with desired international standards that govern the use of force. Add to that mixture the rapidly growing application of artificial intelligence (AI) technology for military purposes, and the need for further discussion becomes critical.

Under conventionally understood wartime rules of engagement, there are two types of allowable targeting: conduct-based and status-based. Attacks within these criteria are usually called signature strikes in the case of the former and personality strikes for the latter. Semi-autonomous weapon systems already exist that target conduct-based threats, such as the Phalanx system that protects U.S. Navy ships from incoming missiles. The Phalanx is programmed to detect, evaluate, and engage an incoming target that meets certain criteria in a fraction of the time it would take a human to respond. Israel's Iron Dome air defence system provides similar protection on land. But what about a weapon system that is not used as a semi-autonomous last line of defence, but rather a fully autonomous offensive hunter-killer?

Most autonomous functions currently in use involve 'weak AI' that simply amounts to a decision tree-like processing of data. 'Strong AI' on the other hand tries to find a solution that fits closest to the programmed objective, and then learns from externally driven experiences. An AI algorithm that is capable of generating its own knowledge tied to a machine that is able to act on that knowledge to unleash lethal force independent of human judgement would circumvent the traditional military leadership chain of command used to direct major combat action.

There is strong resistance, even from within the military community, to fusing strong AI with full lethality. An established targeting methodology known as find-fix-finish fuses intelligence sources with tracking technology to engage a target. Research is ongoing into using advanced processes such as facial recognition technology for drone systems to at least find and fix a targeted individual. At least for the foreseeable future, the 'finish' decision will still be the responsibility of a human decision maker within the kill chain.

Some say there is a moral worth to drones because they are more discriminant, proportional and less destructive than other weapon systems. Others argue that the relative low cost and availability of drone systems make their use more likely than would otherwise be the case if the lives of friendly troops or aircrew were to be risked. While drones have transformed some practices of war, the ethical and legal questions surrounding them relate more to how they are used and not the weapon systems themselves. With the ability to loiter high in the air for up to 24 hours while using multi-spectrum sensors, discrimination should be easy to achieve using a drone. Proportionality would also be possible by simply waiting for the right moment to strike, preferably when innocent civilians are not in the immediate vicinity. The smaller size of drones compared to piloted aircraft has resulted in the development of smaller munitions – some with no explosive warhead at all. This in turn has put a premium on hitting the right target and only the right target. But who (or what) do we want making that decision and launching that missile?

JUST WAR THEORY

Classical Just War thinking had developed in the Western mind from the writings of Aristotle through Cicero and up to Augustine of Hippo. Concerned with maintaining order, the aim of a just war was the restoration of a just peace. For peace to be the aim of war, virtue was required to properly restore a just and tranquil order. Thomas Aquinas would later extend and refine some of these concepts, which would then find their way through the natural law interpretations of Francisco de Vitoria, Hugo Grotius and Francisco Suarez to their present form in international law and treaty conventions. For military personnel, the Law of Armed Conflict and the Rules of Engagement apply those concepts to the realities of the battlefield.

According to Just War thinking, there are three key factors to consider before engaging in a war (*jus ad bellum*): in order to be considered just, a war must be declared by a legitimate authority, be initiated for a just cause and with right intentions. Over the centuries, elements were added to include using force proportionate to the threat, having some prospect for success and the recourse to arms only as a last resort. From the various early definitions of what constituted a just cause, later Thomistic thinkers began to settle on self-defense as the sole legitimate justification for waging war. After witnessing mass killings and genocide in places like Rwanda, late twentieth-century world leaders would add the responsibility to protect (R2P) defenceless civilians as another justifiable reason.

Apart from justification for starting a war, once hostilities have commenced there are several principles that apply to how one fights. For the justification of actions during a war – *jus in bello* – there must still be a continuing assessment of right intention. To this is added the idea of proportionality to avoid wanton destruction, and discrimination in order to distinguish combatants from noncombatants. Armies should only use the amount of force necessary to achieve their military objectives. Deliberately targeting civilians is to be avoided. If unavoidable, the proportionality rule dictates minimizing noncombatant casualties as much as possible. Put simply by Just War thinkers such as Aquinas, the calculus of actions conducted during war must result in doing more good than evil.

DRONES

The use of drones during an armed conflict is ostensibly governed by international law and the military Law of Armed Conflict. Under rules of combat informed by *jus in bello* concepts, inflicting death on an enemy combatant is accepted as a measure not of last resort, but of first choice. On the battlefield, enemy forces are legitimate targets for military attack whether they are chosen by identity or by deed, subject only to necessity and proportionality (Meisels, 2018). Irregular forces that flaunt the laws of war are considered by some to have relinquished the conventional protections accorded to lawful combatants, such as the right not to be targeted in civilian locations or to the status of prisoner of war when captured.

While Geneva Conventions proscribe attacks causing civilian casualties 'excessive in relation to concrete and direct military advantage anticipated,' it outright condemns intentional attacks on non-combatants (Protocol, 1977, 38). Under international humanitarian law that applies during wartime, belligerents have the legal right to harm each other. In times of peace, international human rights law permits killing someone only to prevent an imminent threat to innocent life.

Terrorist groups blur the line between combatant and criminal. They take their war with them wherever they go and avoid the 'heavy quantum of force' that comes with the more organized and intense fighting imagined by Just War adherents (Brunstetter & Braun, 2013, 92). International terrorism defies common classification in the Hague and Geneva conventions and in the United Nations framework, which are more geared toward interstate conflict or armed rebellion within a country. Because terrorist groups often 'hide in plain sight' among civilian populations, they cause noncombatants to be drawn into a conflict and placed in harm's way. One solution to this problem has been to use remotely piloted aircraft that can loiter in an area for long periods of time, maintain surveillance on a suspected target and then strike when the moment is right and the target is isolated.

The most common military terms for what are popularly called drones basically describe the fact that they do not have a pilot onboard. Unmanned Aerial Vehicle or UAV, a term used to describe the actual aircraft, has evolved into the broader Unmanned Aerial System or UAS in order to capture the complexity of a weapon system that involves ground, air and space communication and technological networks that span the globe. The term Remotely Piloted Aircraft or RPA is also used and perhaps more accurate in that it indicates how the aircraft is operated: remotely. However, 'drone' has entered the popular vocabulary as an easy shorthand term. Of the early systems that are still around and prevalent, the two most people are familiar with are the MQ-1 Predator and the MQ-9 Reaper. Many other smaller armed drones – including one-time use 'suicide' drones – have flooded the international market and are produced by a growing number of countries. Some of the smallest can be launched by hand.

Without the required life-support systems of a piloted or crewed aircraft, a UAV is smaller than any conventional fighter jet or bomber. It carries fewer bombs and missiles than a fighter-bomber, and those munitions are themselves smaller. Whereas a dual role fighter-bomber might carry several 900-kilogram (2,000-pound) bombs, drones are limited to a combination of a few 45-kg (100-lb) missiles and 225-kg (500-lb) laser-guided bombs at best. Depending upon the fusing, these smaller munitions can have a significantly reduced blast radius. A relatively new attempt to further limit collateral damage and civilian deaths has resulted in the use of missiles that carry no warhead at all. The AGM-114 Hellfire variant known as the R9X uses sheer kinetic force and six blades to isolate damage to just the intended target. Al-Qaeda leader Ayman al-Zawahiri was killed in Kabul by a drone strike in 2022 using this type of missile.

In wartime and on the battlefield, drone strikes basically follow the same rules as other attacks. Enemy forces are fair game whether it is because of who they are or what they are doing. In areas of active combat between opposing forces, many enemy soldiers are targeted not necessarily because of what they are doing at the moment, but because of who they are. The ethical dilemma of the 'naked soldier' is resolved using the logic of imminence and continuing threat: an unarmed enemy soldier bathing in a river is a legitimate target because of the future threat posed once dressed and rearmed. Soldiers can legally be killed in war anytime – unless they are a prisoner of war, have surrendered or are incapacitated – because they continuously pose a threat (Restrepo, 2019). Also, as a member of a military force, they are considered a part of a collective that threatens the lives of members of another collective (Restrepo, 2019).

JUSTIFYING THE MEANS

Article 51 of the United Nations Charter provides the legal permission for the use of force in self-defense. When the United States was attacked on September 11, 2001, Congress quickly passed the Authorization for the Use of Military Force (AUMF) which gave the president the legal right to pursue lethal action against terrorist groups that posed a threat to Americans. Many of these groups were operating out of territories that were ungoverned or poorly governed, mostly in remote and difficult to reach areas. The MQ-1 Predator was already being developed to deploy munitions in addition to its reconnaissance capabilities; after 9-11, that became a priority. With the ability to loiter over hostile territory for long periods of time, maintain surveillance on a target and then strike with precision weapons, it seemed like the perfect new system for this new type of warfare against a new type of global enemy.

Drones were used by the administration of President George W. Bush in tandem with conventional and special forces to prosecute the war on terrorism. Under President Barack Obama, their use continued in an upward trend. President Obama's National Security Council counterterrorism advisor (later CIA director) John Brennan stated that international law did not ban drones or prohibit force outside of the battlefield, especially when the country where a terrorist was located consents or is unable or unwilling to act against the threat. A Catholic and self-described 'adherent of just-war theory,' he was setting the stage for the next year, when President Obama publicly explained his position on using drone warfare to combat terrorism and followed up with presidential policy guidance that would outline specific procedures (Braun, 2020, 584).

In a speech at the National Defense University in May of 2013, President Obama laid out the U.S. policy on drone strikes:

> [T]he United States has taken lethal, targeted action against al Qaeda and its associated forces, including with remotely piloted aircraft commonly referred to as drones. Under domestic law and international law, the United States is at war. This is a just war, waged proportionally, as a last resort, and in self-defense. America does not take strikes to punish individuals; we act against terrorists who pose a continuing and imminent threat … when there are no other governments capable of effectively addressing the threat … [and with] near certainty that no civilians will be killed or injured.
>
> (Obama, 2013)

In his speech, President Obama explained that targeting terrorists in an armed conflict 'as a preventive, rather than a punitive, measure is a legitimate defensive act … subject to military necessity and proportionality' (Meisels, 2018, 6-7). In this way, he hoped to avoid confusing a targeted killing of a terrorist in line with international humanitarian law (wartime) with a police action, which would have fallen under international human rights law (peacetime). In a corresponding Presidential Policy Guidance on 'Procedures for Approving Direct Action Against Terrorist Targets Located Outside of the United States and Areas of Active Hostilities,' Obama outlined three key elements for approving lethal action against terrorist targets outside of hot battlefields: they must present a continuing, imminent threat; capture is not feasible; and there must be 'near certainty' that the target will be hit and that non-combatants would not be harmed or killed.

In 2018, President Donald Trump loosened some of these restrictions by delegating strike authority further down the chain of command, eliminating the requirement for a targeted individual to pose an imminent threat, changing the need for 'near certainty' to 'reasonable certainty' and reducing policymaker involvement and oversight. Immediately upon taking office in January of 2021, President Joseph Biden issued stopgap measures to reverse these policies. Returned were National Command Authority approval for target selection, the imminence of a threat and a near certainty of target identity and the absence of non-combatants.

When President Obama had explained the use of drones as part of 'a just war, waged proportionally, as a last resort, and in self-defense,' he channelled classic Just War thinking in order to defend his policy. Both then and now, world leaders assess that some justification is required when considering how to start, wage and end a war. However, the public's ethical views on drone warfare can sometimes be difficult to discern. In 2015, the United States Conference of Catholic Bishops sent a letter to U.S. government officials, including the National Security Advisor Susan Rice:

> Using armed drones for targeted killings should be limited to those areas of intense, active, and protracted conflict where there have been declarations of war, where there is multilateral agreement that such action is needed to counter extreme violence perpetrated on non-combatants, and when the target is a combatant who is likely soon to launch an attack. Armed drones may be used outside of areas of open and protracted fighting if it is determined that the person targeted poses an immediate threat, if the use of lethal force is proportionate and there is no other means to prevent the threat to life, and if civilian casualties can be avoided as much as possible.
>
> (Cantu, 2015)

Avoiding civilian casualties 'as much as possible' is a far cry from the demands for peace and justice a few decades earlier from religious groups and non-governmental organizations. Perhaps the degree of discrimination and proportionality attainable using drones *is* as just as war gets.

JUST WAR THEORY MEETS THE TWENTY-FIRST CENTURY

Some Just War theorists are calling for a new alternative to *jus in bello* principles due to issues raised by drone warfare. According to Megan Braun and Daniel Brunstetter, 'Achieving ethical drone operations requires a more calibrated moral framework' (Braun & Brunstetter, 2013, 320). They propose a more nuanced approach which acknowledges the nature of *jus ad vim* or force-short-of-war. They suggest that strikes against elusive targets, hidden among civilians and far from active battlefields, demand stricter criteria for proportionality and discrimination (Brunstetter & Braun, 2013). With a stronger correlation between use of force and intended effects, there needs to be a more vigilant and continuous assessment of lethal action. 'Whereas in war, principles such as just cause and last resort need only be satisfied at the outset of a conflict, *jus ad vim* requires that they be continually reassessed in advance of each use of force…' (Braun & Brunstetter, 2013, 317). As a justification of force application short of all-out war, *jus ad vim* allows for a more nuanced and measured approach to targeting.

Enemark offers one more subtle distinction to classical Just War theory by pointing to the new reality of war and specifically the global war on terrorism. With *vis perpetua* (perpetual force), he draws from Yee-Kuang Heng's thoughts on the transformation of war to suggest that certain threats can never be completely eradicated; they can only be managed (Enemark, 2014). This type of forever war becomes an endless process of risk management where victory is measured by the number of foiled or failed terrorist attacks. Drone actions in this case are justified by the resulting reduction of planned and actual threats from terrorist groups.

New research into the impact of drone operations on civilians has revealed side effects that go beyond immediate casualties. Harm is not simply measured in injuries and deaths, but also in human rights concerns over the psychological impact on communities (Brunstetter & Braun, 2013). Feelings of insecurity and vulnerability can have long-lasting effects on civilian populations. To temper the conclusion of this study, some defenders of drone strikes counter that, 'the fact of civilians *feeling* threatened by violence does not equate to the moral wrong of deliberately *threatening* people who are understood to be civilians' (Enemark, 2022, 312). Others claim that in the case of drones, psychological harm is justifiable to the extent that it is incurred sincerely as an undesirable side-effect of the war on terror (Meisels, 2018). Still, the fact that we know about the second and third order impacts of drone missions on non-combatant populations means that it must be considered in any cost-benefit calculation. Psychological scars that remain long after conflicts are over can harm efforts at long-term reconciliation and peacebuilding – once again, peace properly being the just aim of a just war.

The ethical use of drones *can* ensure discrimination and proportionality, with fewer civilian casualties. A drone operator can exercise a higher degree of restraint when faced with ambiguous potential threats than a soldier who could be killed due to a miscalculation in the heat of battle (Schulzke, 2016). A warfighter in a combat zone, bound by the Law of Armed Conflict, international conventions and rules of engagement, is concerned with the here and now – *here* being a few feet away and *now* being a matter of split seconds. Still, for the drone operator, it is said that contact with the enemy is not measured in thousands of miles from the target so much as several inches from the computer screen. However it is measured, drone warfare should not be allowed to depart from the ethical constraints placed upon all other means of employing destructive force.

Whether at 10 feet or 10,000 miles, the ethical judgment needed to determine whether to take the shot is one and the same. The elongated spatial range and extended timeframe offered by drones make deliberate targeting possible. With the ability to access remote or dangerous regions quickly and with low risk, drones make dynamic targeting effective. Either way, when considering the goal of effects-based targeting and weaponeering, an unmanned aerial system offers advantages in accuracy, discretion, and proportionality. Lower yield bombs or missiles carried by those drones limit the destructive force that is unleashed. Finally, artificial intelligence that can enhance the capability to discriminate and rapidly calculate proportionality may result in a more just application of lethal force. Whether those systems are used in a manner consistent with Just War principles is a matter of policy, which can sometimes change.

ARTIFICIAL INTELLIGENCE AND THE DECISION LOOP

For weapon systems such as drones, it is the manner in which they are used more than the weapon itself that determines whether it can be considered just or not. But what if the weapon itself had the ability to decide when and what it can target and then act on that decision? Enter the realm of lethal autonomous weapon systems or LAWS. Noreen Hertzfeld emphasizes the importance of addressing the ethics of these systems: 'Just as the advent of nuclear weapons caused twentieth-century theologians to reevaluate the justness of war, the advent of lethal autonomous weapon systems presents an even more urgent call to twenty-first century theologians to do likewise' (Hertzfeld, 2022, 70).

At least for the near future, it seems as though most military leaders are reluctant to turn unfettered control of large-scale offensive weapon systems over to computers. The U.S. Air Force philosophy of centralized control and decentralized execution to conduct air operations is based on human leadership as the connecting tissue of multi-domain warfare. AI may improve the speed and quality of battlefield assessments – and even suggest courses of action – but legal command authority rests only with duly appointed officers. All that any good commander could ask for from a machine is to provide better situational awareness than the enemy has available. Getting inside an adversary's decision loop is what fast computing is most prized for.

In the current conflict in Gaza, the Israeli military is using an AI-enabled target-generating system called Gospel to rapidly assess Hamas weapons and force dispositions and produce a list of targets in a fraction of the time that it would take humans. However, the tactical advantage of generating shock effect by hitting faster, harder and more often in order to overwhelm an adversary is negated at the strategic level by the international outcry over the disproportional devastation inflicted upon the civilian population that Hamas militants use as human shields. Although the Gospel system is generating the targets based upon the algorithms it is programmed to follow, there is a human commander in the loop who chooses to strike them.

With far-reaching repercussions, the stakes are raised for ensuring accurate methods of distinction as well as ethical calculations of proportionality. According to Hertzfeld, the necessary degree of discrimination 'requires a high level of context sensitivity, one that would be complex to program' (Hertzfeld, 2022, 78). With the difficulty of programmers to envision all possible scenarios that the system may encounter or how it would react, universal moral values would be too problematic to encode into any logic algorithm. The system would have to depend on what de

Swarte called context ethics. 'Each ethical rule depends on the context in which ... it is used' (de Swarte et al., 2019, 291). So unlike chess, which has fixed rules for moving pieces, warfare presents an unlimited number of situations obscured by fog, friction and the uncertainty of battle. And war certainly is not a game of chess.

With stronger AI comes increased unpredictability. The same way that early AI systems that were designed to play chess came up with moves that experts would never had thought of, fully autonomous AI-enabled weapon systems may develop tactics that were not foreseen by designers. And once a missile comes off the rail, there can be very little time or ability to stop it in a human-out-of-the-loop system. The machine's decision and action would be irreversible. The farther up the chain of command this weapon is inserted, the more profound would be the consequences of its choices.

Several countries are accelerating research into AI-enabled weapon systems. Depending upon the country, more or less rigorous standards for testing and evaluation work to minimize biases which could lead to unintended consequences. In addition to the high number of simulations pitting the system against itself, actual operational tests are crucial. In 2020, an AI system called ARTUμ shared control of a U.S. Air Force U-2 reconnaissance plane's radar, allowing the pilot to concentrate on flying while the computer determined how to toggle the radar between searching for enemy ground missile launchers and detecting airborne threats to the aircraft (Tegler, 2020). This type of human-machine teaming may be the norm for most systems that become operational in the next few years. Not only will it take time to build trust between the human and the machine, it will perhaps take even longer to adapt the human warrior spirit to the new partnership. Beyond the testing stage, some forms of loitering munitions have already been fielded that can be programmed to fly a particular pattern but then react to an enemy action and launch a missile or even guide itself to engage and destroy the threat. In this reactive, defensive role they are similar to the Phalanx close-in system and the Iron Dome air defence system, although loitering munitions project that defensive role much further outward.

Fully autonomous AI-enabled weapons can make decisions on their own, at an incredibly fast speed and with amazing precision. They can also make unexpected and unfathomable choices.

In war, it is already difficult to anticipate the actions of the enemy. Opposing commanders may be reasoning from a completely foreign set of principles. The soldiers operating weapons systems may make unpredictable split-second decisions that run counter to military doctrine. And now, as Hertzfeld has pointed out, 'autonomous weapons add to this confusion by introducing a new potential actor ... namely, the weapon itself' (Hertzfeld, 2022, 72). The human element of conflict already makes Just War principles difficult to apply, imagine now having to consider the weapon as a 'responsible' actor. Whose virtue is at stake when a machine commits an unjust act?

Where would the responsibility fall for ensuring civilians are not targeted or that proportionality considerations properly weigh legitimate military objectives against unnecessary suffering? Human in-the-loop systems require a person to manually perform some action in order for a process to continue. Human on-the-loop systems have someone overseeing operations with the ability to pause or stop a process from continuing. Human out-of-the-loop weapon systems are completely autonomous and may take lethal action on their own. A landmine is a simple example of an autonomous weapon that once set needs no further human directions in order to explode. However, it does not have the ability to 'decide' whether to detonate – it just does.

Senior military leaders are very reluctant to employ weapons whose effects are uncontrollable. National leaders in the past agreed, for instance when they banned chemical and biological weapons. Once these types of weapons are unleashed, there is no telling what the extent of suffering and damage might be, or when it would stop. Although AI-enabled weapons may have the potential to demonstrate unpredictable behavior of an unimaginable nature leading to unintended consequences, the risks may be mitigated by restricting their autonomy. The amount of risk should determine the degree of machine freedom. The greater the risk, the more human control would be needed. Conversely, as Rosalind Picard pointed out, 'The greater the freedom of a machine, the more it will need moral standards' (Picard, 1997, 19).

CONCLUSION

The United States, in justifying its reluctance to supply more advanced weapons to Ukraine during the opening stages of the current Russian invasion, voiced its concern that a rapid escalation may touch off World War III. Drones have so far helped keep the air war in Ukraine just barely below the boiling point of regional escalation. But is fighting this type of war simply sustaining a version of what Pope Francis had called 'a world war fought piecemeal' (Francis, 2017). Waging an AI-enabled drone war in heavily populated civilian areas with no end in sight cannot possibly lead to a post-war tranquil order or just peace. A weapon system based on a remotely piloted aircraft may be ethically neutral. Still, how it is used must be weighed against Just War concepts and other considerations. A remotely controlled or autonomous weapon removes friendly forces from immediate danger and a precision munition reduces the risk of non-combatant deaths. These factors can increase the attractiveness of using force short of last resort. In this regard, Tamar Meisels echoes Michael Walzer's warning that, 'The capacity for riskless warfare makes drones dangerously tempting' (Meisels, 2018, 12). In a conflict, putting distance between you and your enemy is not necessarily a bad thing. When that distance is closed and your adversary is bearing down on you with speed and force, having an autonomous weapon system that can engage the in-coming threat simply augments the act of self-defense. The question from the *jus in bello* perspective becomes this: do these systems help to distinguish civilians, reduce unnecessary casualties and avoid escalation which would cause more suffering? Are they doing more good than harm?

AI technology that can rapidly distinguish combatants from noncombatants and a drone that provides loitering standoff capability are each compatible with Just War principles. The proportionality of using small precision-guided munitions – some with no warhead at all – launched from a drone and thus reducing or eliminating civilian casualties, is also compatible.

Decisive action in warfare oftentimes involves resolving ethical conflicts that appear to be unsolvable in the short time it takes to react to a threat or opportunity. There may be no discernible clear-cut solution; only some outcomes that are more desirable than others. Seemingly rule-abiding smart machines are great – until they are not. When AI faces a new situation that it hasn't encountered before, there is no telling what the result might be. Is the risk worth it?

REFERENCES

Archambault, E. (2020). A Good Guy with a Drone: On the Ethics of Drone Warfare. *Contemporary Political Theory*, 19(Suppl 3), 169–175. https://doi.org/10.1057/s41296-019-00328-w

Braun, C. N. (2020). The Catholic Presumption Against War Revisited. *International Relations*, 34(4), 583–602. https://doi.org/10.1177/0047117819879486

Braun, M. & Brunstetter, D. R. (2013). Rethinking the Criterion for Assessing Cia-targeted Killings: Drones, Proportionality and Jus Ad Vim. *Journal of Military Ethics*, 12(4), 304–324. https://doi.org/10.1080/15027570.2013.869390

Brunstetter, D. R. & Braun, M. (2013). From *Jus Ad Bellum* to *Jus Ad Vim*: Recalibrating Our Understanding of the Moral Use of Force. *Ethics & International Affairs*, 27(1), 87–106. https://doi.org/10.1017/S0892679412000792

Burkhardt, T. (2015). Justified Drone Strikes Are Predicated on R2P Norms. *International Journal of Applied Philosophy*, 29(2), 167–176.

Cahill, L. S. (2015). Just War and the Gospel. In T. Winright & L. Johnson (Eds.), *Can War Be Just in the 21st Century?: Ethicists Engage the Tradition* (pp. 20–27). Orbis Books.

Cahill, L. S. (2019). Just War, Pacifism, Just Peace, and Peacebuilding. *Theological Studies*, 80(1), 169–185. https://doi.org/10.1177/0040563918819808

Cantu, O. (2015, May 11). *Letter from the Committee on International Justice and Peace to Ambassador Susan Rice*. United States Conference of Catholic Bishops. https://www.usccb.org/issues-and-action/human-life-and-dignity/war-and-peace/arms-trade/upload/letter-to-nsa-rice-from-bishop-cantu-re-drones-policy-framework-2015-05-11.pdf

Corn, G. S. (2009). What Law Applies to the War on Terror? In PALCI EBSCO books, M. W. Lewis & G. S. Corn (Eds.), *The War on Terror and the Laws of War: A Military Perspective* (pp. 1–36). Oxford University Press.

Corn, G. S. & and Jensen, E. T. (2008). Untying the Gordian Knot: A Proposal for Determining Applicability of the Laws of War to the War on Terror. *Temple Law Review*, 81(3), 787–830.

de Swarte, T., Boufous, O. & Escalle, P. (2019). Artificial Intelligence, Ethics and Human Value: The Cases of Military Drones and Companion Robots. *Artificial Life and Robotics*, 24, 291–296. https://doi.org/10.1007/s10015-019-00525-1

Douglas, M. (2003). Changing the Rules: Just War Theory in the Twenty-first Century. *Theology Today*, 59(4), 529–545.

Enemark, C. (2014). Drones, Risk, and Perpetual Force. *Ethics & International Affairs*, 28(3), 365–381. https://doi.org/10.1017/S0892679414000446

Enemark, C. (2022). The Enduring Problem of 'Grey' Drone Violence. *European Journal of International Security*, 7(3), 304–321. https://doi.org/10.1017/eis.2021.24

Francis. (2017, January 1). *Nonviolence: A Style of Politics for Peace*. https://www.vatican.va/content/francesco/en/messages/peace/documents/papa-francesco_20161208_messaggio-l-giornata-mondiale-pace-2017.html

Hertzfeld, N. (2022). Can Lethal Autonomous Weapons Be Just? *Journal of Moral Theology*, 11(SI1), 70–86.

Himes, K. R. (2004). Intervention, Just War, and U.S. National Security. *Theological Studies*, 65(1), 141–157.

Hogan, M. M. (2012). Just War: A Catholic Perspective: *Cui Non Videtur Causa Justa*? *American Journal of Economics and Sociology*, 71(4), 956–965. https://doi.org/10.1111/j.1536-7150.2012.00856.x

Meisels, T. (2018). Targeted Killing with Drones? Old Arguments, New Technologies. *Philosophy and Society*, 29(1), 3–16. https://doi.org/10.2298/FID1801003M

Obama, B. (2013, May 23). *Remarks by the President at the National Defense University*. Transcript of speech delivered at the National Defense University, Fort McNair, Washington, DC. https://obamawhitehouse.archives.gov/the-pressoffice/2013/05/23/remarks-president-national-defense-university

Picard, R. W. (1997). *Affective Computing*. The MIT Press. https://search-ebscohost-com.ezp1.villanova.edu/login.aspx?direct=true&db=e000xna&AN=397

Protocol Additional to the Geneva Conventions of 12 August 1949, Article 51, 5(b) (1977). https://ihl-databases.icrc.org/en/ihl-treaties/api-1977/article-51?activeTab=undefined

Provost-Smith, P. (2013). A Drone's Eye View: Global Anti-Terrorism and the Existential Crisis of Just War Theory. *Journal for Cultural & Religious Theory*, 12(3), 84–119.

Restrepo, D. (2019). Naked Soldiers, Naked Terrorists, and the Justifiability of Drone Warfare. *Social Theory and Practice*, 45(1), 103. https://doi.org/10.2307/45218913

Riza, M. S. (2014). Two-Dimensional Warfare: Combatants, Warriors, and Our Post-Predator Collective Experience. *Journal of Military Ethics*, 13(3), 257–273. https://doi.org/10.1080/15027570.2014.976475

Schulzke, M. (2016). Rethinking Military Virtue Ethics in an Age of Unmanned Weapons. *Journal of Military Ethics*, 15(3), 187–204. https://doi.org/10.1080/15027570.2016.1257851

Stiltner, B. (2015). A Taste of Armageddon: When Warring is Done by Drones and Robots. In T. Winright & L. Johnson (Eds.), *Can War Be Just in the 21st Century?: Ethicists Engage the Tradition* (28–38). Orbis Books.

Stimson Center. (2018). *An Action Plan on U.S. Drone Policy: Recommendations for the Trump Administration*. https://www.stimson.org/sites/default/files/file-attachments/Stimson%20Action%20Plan%20on%20US%20Drone%20Policy.pdf

Tegler, E. (2020, December 16), An AI-Pilot Called 'ARTUμ' Just Took Command of a U-2's Sensor Systems on a Recon Mission, *Forbes*, https://www.forbes.com/sites/erictegler/2020/12/16/an-ai-co-pilot-called-artujust-took-command-of-a-u-2s-sensor-systems-on-a-reconnaissance-mission/?sh=6d58ceab61f0

Walzer, M. (2013, January 11). Targeted Killing and Drone Warfare. *Dissent*. http://www.dissentmagazine.org/online_articles/targeted-killing-and-drone-warfare

6 On the Morality of Fully-Autonomous Armed Drones

Christian Nikolaus Braun and Bernhard Koch

There is arguably no other weapons system in recent times that rivals the critical attention that armed uninhabited aerial vehicles (UAVs), commonly referred to as armed drones, have received from ethicists. Scholars of various backgrounds have brought to bear their expertise on a weapons system whose role in contemporary warfare seems to be ever increasing (see, e.g. Braun, 2021a; Brunstetter & Braun, 2011; Buhta et al., 2016; Enemark, 2023; Himes, 2016; Kaag & Kreps, 2014; Koch, 2016, 2019a). In fact, scholarship on armed drones is already referring to the 'three waves' of drone research that have investigated the proliferation, the effectiveness, and the legal and moral implications of armed drones (Lushenko, Bose, & Maley, 2021). While early drone research concentrated on the then novel weaponising of UAV systems, critical attention is now increasingly turning towards the prospect of fully autonomous drones.

In fact, what seemed like science-fiction just a few years ago has turned into one of the most concerning ethical and legal questions of our time (see, e.g. Bode & Huells, 2022; Baker, 2022; Galliott, MacIntosh, & Ohlin, 2021; Grünwald & Kehl, 2020; O'Connell, 2023). As the ongoing war against Ukraine demonstrates, the international community is facing an epoch-defining moment. One of the characteristics of this war has been the large-scale use of armed drones. The systems employed in Ukraine increasingly rely on AI, and there is emerging evidence that 'fully autonomous' combat operations have been carried out already (Hambling, 2023). In fact, the technology to let armed drones function fully autonomously has existed for some time (Bode & Watts, 2023). In consequence, for the first time in human history, humankind is facing the prospect of hostilities being conducted in the absence of human agency.

In this chapter, we seek to make moral sense of the advent of potentially fully autonomous armed drones, which would resemble a so-called AWS (Autonomous Weapons System). When we speak of AWS, we have in mind the broad definition provided by the ICRC (2021): 'Weapons that select and apply force to targets without human intervention.' Our reflections are rooted in the Catholic tradition. In particular, we focus on the concept of human dignity that is at the very heart of Catholic social thought. From the Church Fathers onwards, the Catholic Church has been what in modern terminology could be described as a norm entrepreneur for Western

DOI: 10.1201/9781003440253-6

thinking about war and peace. Until the present day, the Church continues to make its voice heard in the debate about the rights and wrongs of war. Commonly, the current Catholic take on the ethics of war and peace is summarised as starting from a 'presumption against war'(National Conference of Catholic Bishops, 1983). The main rationale of this starting point is to make the resort to armed force exceptional. Only if a number of restrictive criteria have been met can the resort to armed force considered to be justifiable (see, e.g. Reichberg, 2002; Hrynkow & Power, 2019; Braun, 2020). Most recently, the late Pope Francis further accentuated this conversation when he (Pope Francis, 2022a) rejected the idea that there can ever be a 'just war' while, at the same time, affirming Ukraine's right to self-defence against Russian aggression (Pope Francis, 2022b).

The general outlook on the ethics of war and peace sketched above also determines the Catholic position on specific weapons systems. For example, much ink was spilled during the Cold War on the Catholic 'interim position' on nuclear weapons that ruled out the use of such weapons but reluctantly accepted possession as a means of keeping a negative peace (see, e.g. Hollenbach, 1982; Reid, 1986). Pope Francis's (2019) recent rejection of both use and possession of nuclear weapons as 'immoral' has revived the Catholic debate about this type of weapon (see Christiansen & Sargent, 2023; Braun, 2024a, 2024b). Armed drones provide another instance where the Catholic Church has made its voice heard. Important voices within the Church, including US (United States Conference of Catholic Bishops, 2013, 2015) and German (Deutsche Kommission *Justitia et Pax* & Katholische Militärseelsorge, 2013) bishops, have commented on the morality of their use in the context of the targeted killings of alleged terrorists in the so-called war on terror. These statements have been highly critical of armed drones, and especially particular practices for which they have been used, and thus reflect the general presumption against war that informs the contemporary Catholic position. However, their reservations about remotely piloted drones notwithstanding, the bishops have not come out as seeing such systems as *mala in se*, an evil in themselves. Facing the prospect of fully autonomous armed drones, the Catholic position seems to be more critical still, arguably approaching the *mala in se* category, although it has not yet been used explicitly by major voices within the Church. Pope Francis (2020) warned that AWS 'irreversibly alter the nature of warfare, detaching it further from human agency.' Francis's permanent observer to the United Nations employed an even stronger wording, arguing that AWS 'will provide the capacity of altering irreversibly the nature of warfare, becoming even more inhumane, putting in question the humanity of our societies'(Jurkovič, 2018).

In terms of outline, it is our ambition to achieve two objectives. Firstly, we trace the human dignity element of the Catholic argument on remotely piloted drones that was made during the second Obama administration. We then turn to the Catholic argument on AWS, imagined in the form of a fully autonomous armed drone. Acknowledging that the former type of drone also receives AI input of various forms, for us the crucial moral difference between the two types of drones are the moments of target selection and execution. While for a remotely piloted drone these two functions are still carried out by a human operator, a fully autonomous drone will perform these functions without human agency involved. Secondly, we take the critical attitude toward armed drones that has been put forward by Church leaders as our starting point to reflect upon the question of whether such systems do in fact violate human dignity. While there is an increasing engagement with AWS on the side of major voices within the Church, these remarks have not been very elaborate to date. In this chapter, we seek to provide a more detailed engagement with the human dignity critique that is grounded in the Catholic tradition. Armed drones, we hold, through the distinction between remotely-piloted and fully autonomous manifestations, provide a rich foundation to explore if the usage of AWS should indeed be proscribed as *mala in se*. In so doing, we also intend to facilitate an exchange between the ethical debates in the English-speaking world and the conversations in German-speaking countries, which to the most part have unfolded independently of each other. We find this state of affairs unfortunate and, by drawing on the two bodies of literature for our own argument, seek to make a modest contribution toward a meaningful exchange on the question of AWS.

DRONES AND HUMAN DIGNITY

At the outset, before turning to the human dignity-based arguments about two basic types of drone technology that have been put forward by major voices within the Church, it is important to note that there is no single authoritative position on these matters within Catholicism. In forming one's own position on questions of war and peace, such as drone technology, ethicists need to weigh the relevant facts, which may lead them to different conclusions. When that is the case, it makes more sense to speak of '*a* Catholic view than *the* Catholic view'(McKenna, 1960, p. 648) on a specific moral issue. In his role as prefect for the Congregation of the Doctrine of the Faith the then Cardinal Joseph Ratzinger (2004), the later Pope Benedict XVI, clarified the question of potential disagreements on ethical issues as follows:

> Not all moral issues have the same moral weight as abortion and euthanasia. For example, if a Catholic were to be at odds with the Holy Father on the application of capital punishment or on the decision to wage war, he would not for that reason be considered unworthy to present himself to receive Holy Communion. While the Church exhorts civil authorities to seek peace, not war, and to exercise discretion and mercy in imposing punishment on criminals, it may still be permissible to take up arms to repel an aggressor or to have recourse to capital punishment. There may be legitimate diversity of opinion even among Catholics about waging war and applying the death penalty, but not however with regard to abortion and euthanasia.

Having provided this contextualisation, which should be kept in mind when we, as scholars working in the Catholic tradition, make our own human dignity-based arguments on AWS, the chapter now turns to the importance of the concept itself. Indeed, human dignity is considered 'the foundation of all Catholic Social Teaching'(Canadian Catholic Organization for Development and Peace, n.d.). The *Compendium of the Social Doctrine of the Church* provides the following explanation about the role of human dignity in a just society:

> A just society can become a reality only when it is based on the respect of the transcendent dignity of the human person. The person represents the ultimate end of society, by which it is ordered to the person: 'Hence, the social order and its development must invariably work to the benefit of the human person, since the order of things is to be subordinate to the order of persons, and not the other way around.'
>
> (Pontifical Council of Justice and Peace, 2004, §132)

Emphasising the role of human dignity for a just society it needs to be noted that its role is not limited to the national level, but extends to international society, too. In other words, human dignity is also of fundamental importance in international affairs, including the conduct of hostilities. But human dignity can be understood in different ways; a Christian-theological understanding starts with the creatureliness of man by God, a Kantian understanding with the moral autonomy of every human subject (see Koch, 2024). Other possibilities are conceivable.

Turning to the Catholic contribution to the debate about remotely-piloted armed drones and human dignity, this aspect was only one area of concern in the high-profile statements published by US and German bishops (see Braun, 2020). The relevant passages in the letter signed by US Bishop Oscar Cantú (United States Conference of Catholic Bishops, 2015) had in mind the consequences of drone use on both the groups at the receiving end of the military force unleashed by them, and the impact of their use on their human operators:

> Drones provoke anxiety among populations where there are targets, inflicting psychological damage on innocent civilians who live in constant fear they may be hurt or killed and listed as 'collateral damage.'... In addition, drone operators, who spend long hours observing their targets and seeing up close the damage they inflict, are likely to suffer high levels of stress.

Furthermore, the 2015 letter briefly engaged with the then emerging concern about fully autonomous drones as the next phase in the technological development of these weapons systems. In a paragraph headlined 'Humans Need to Remain in Control of Armed Drones,' Bishop Cantú raised the following moral questions:

> In the face of efforts to create fully autonomous weapon systems, the Holy See has said that humans should always be involved in the use of armed drones, because 'pre-programmed, automated technical systems' lack the ability 'to make moral judgments over life and death, to respect human rights and to comply with the principle of humanity.'... The use of fully autonomous killer drones raises serious questions of moral accountability: In wrongful deaths, who is morally responsible? The persons who deployed the drone? Those who programmed it? Those who manufactured it? Who is morally culpable in wrongful civilian deaths from an autonomous drone?

Although not explicitly invoking human dignity, the concept clearly undergirded parts of his reasoning. Interestingly, German bishops Stephan Ackermann and Franz-Josef Overbeck (Deutsche Kommission *Justitia et Pax* & Katholische Militärseelsorge, 2013), in a 2013 press release, included a brief reference to fully-autonomous drones that, likewise, alluded to human dignity: 'We turn against the use of drones as instruments of execution. This applies even more when, as is already being discussed today, drones are expected to take the decision to kill 'autonomously.' The killing of a human being must not be triggered by a mechanism.'

Five years after the German bishops, when the advent of AWS had come closer and the debate about their morality was receiving increased attention, the Holy See's Permanent Observer to the United Nations and Other International Organisations in Geneva, Archbishop Ivan Jurkovič, highlighted the human dignity aspect in a statement at the 2018 Group of Governmental Experts on Lethal Autonomous Weapons Systems. In particular, the archbishop (Jurkovič, 2018) proposed the principle 'anthropological non-contradiction' as 'the common point of reference' for the discussion about AWS. Crucially, underpinning this principle is the concept of human dignity: 'In other words, any technology, including autonomous weapons, in order to be acceptable, must be compatible and consistent with the proper conception of the human person, the very foundation of law and ethics'(Jurkovič, 2018). In his statement, the archbishop made an effort to contrast the decision-making of a machine with that of a human being, which he traced to human dignity:

> This rationale ignores the fact that for a machine, a human person, just like everything else, is merely an interchangeable set of data among others. Now, this is precisely the 'inhuman'contradiction: the delegation of powers to autonomous systems puts us on the path of negation, oblivion and contempt for the essential characteristics unique to the human persons and to soldierly virtues.
>
> (Jurkovič, 2018)

Furthermore, according to Jurkovič (2018), allowing the advent of AWS would send the world community on a dangerous trajectory that might result in more conflict: 'A world in which autonomous systems are left to manage, rigidly or randomly, fundamental questions related to the lives of human beings and nations, would lead us imperceptibly to dehumanization and to a weakening of the bonds of a true and lasting fraternity of the human family.'

As a result, the pope's envoy pronounced profound reservations against the emergence of AWS that put a premium on what he saw as inherent challenges of these weapons systems to the concept of human dignity (see Braun, 2021b). In doing so, he anticipated one of the main arguments against AWS, which has also featured prominently in the fully secular academic debate, mainly in the argument about AWS as undermining the so-called moral equality of combatants (see, e.g. Skerker, Purves, & Jenkins, 2020). In what follows, we will engage with the archbishop's human dignity-based argument against AWS. Grounded in the Catholic tradition, we will provide an in-depth exploration of human dignity vis-à-vis AWS. As a result, we hope to advance the debate about the ethics of AWS, which has not yet received a detailed Catholic assessment.

FROM REMOTELY-PILOTED DRONES TO FULLY-AUTONOMOUS DRONES

The early ethical debates about autonomous weapons systems were characterised by images that were obviously related to movie scripts. Not surprisingly, early trials of semi-autonomous devices such as the SAMSUNG SGR-A1, used by South Korea at its border with North Korea for border protection purposes about a decade ago, caused considerable unease (Claudi, 2015). The frightening fantasies captured in the 'Blade Runner' or 'Terminator' films pointed to anthropomorphic robots, often indistinguishable from humans on the outside and living amongst them, and yet they were lacking crucial human abilities such as compassion. Conversely, their human appearance of all things could be so perfectly designed that it inspired human empathy. However, such scenarios are not realistic, at least for the foreseeable future, vis-à-vis the subject of this chapter. Rather than cyborgs that are almost indistinguishable from humans, we need to grapple with the ethics of AI-enabled weapons systems with various forms of autonomy, one form of which will be AI-enabled armed drones.

As the above engagement with particular Catholic arguments has demonstrated, the use of remotely-piloted armed drones has been a reality for a significant amount of time now. Armed drones experienced their first 'heyday' during the first administration of US President Barack Obama (2009–2013) in operations in the Afghan-Pakistani border region. They were primarily used for so-called 'targeted killings,' in which individually selected persons were targeted. However, even back then there were reports that, in addition to targeted killings, so-called 'signature strikes' were also carried out, in which the target was no longer known individually (and by name), but only classified as dangerous or hostile based on patterns of behaviour. Importantly, signature strikes are enabled by drone technology because drones combine combat and high-level reconnaissance capabilities. However, the sensor technology of drones collects such an abundance of data that it cannot be processed by humans on their own. This is where the role of software comes into play. Therefore, the ethical discussion today needs to focus on the software behind drones and, as a result, on the area of machine learning – a form of Artificial Intelligence. This is why the term 'autonomy in weapon systems' is now used (see Heinrich Boell Foundation, 2018), as it is no longer a question of whether software enables certain controls (e.g. landing a drone on an aircraft carrier) without human intervention, but rather the question of which functions should not be left to the software (e.g. target selection and decision to attack) (see Davies et al. 2023).

The first important ethical question that arises when large amounts of data are processed by algorithms is whether artificial intelligence extends or restricts human decision-making. The German Ethics Council has made this criterion the central aspect of its opinion on the use of AI:

> The basic question and benchmark for the ethical evaluation of technological developments and their use in various contexts presented here is whether the delegation of activities to machines – up to and including possible replacement – expands or diminishes the conditions for responsible behaviour and human authorship.
>
> (German Ethics Council, 2023, p. 79)

However, the topic of autonomous weapon systems is not directly addressed in the statement. Initially, one would assume that a more comprehensive overview of the situation ('Lagebild'), i.e. one that contains more data, expands human freedom of action and in this sense also provides a 'better' overview. Conversely, ignorance is a deficiency in our ability to act; those who know more can make better decisions. At first look, this is certainly true. But in the context of automated data processing, we can also observe the opposite movement: In order for us to remain capable of making decisions – and thus capable of acting – in the face of an overabundance of data, we expect a technical decision-making aid (in particular through pre-sorting). However, such a decision-making aid becomes the 'better'the more specific it becomes. If the technology then suggests a decision, we usually no longer have any good reasons not to take the suggested action. This connection could already be demonstrated with simple chess computers (Koch, 2023), but similar problems arise in other fields, such as medical diagnostics: on what grounds should we distrust a powerful artificial intelligence when it diagnoses certain tumour diseases, for example? After all, it was assumed that it would significantly surpass human capabilities in terms of data acquisition and processing. However, the human actor no longer feels empowered to act through such assistance, but merely degraded to the execution organ of technological functionality. Such forms of disempowerment through technology are relevant to human dignity and also affect the use of weapons systems in the transition from in-the-loop-systems (such as remotely-piloted armed drones) to on-the-loop-systems, finally to out-of-the-loop-systems (such a fully-autonomous drones where both target selection and execution are carried out by AI). It represents a moment of what Pope Francis has already described in 'Laudato Sí'– following Romano Guardini (Guardini, 2024) – as a 'technocratic paradigm'(Pope Francis, 2015, § 106–114).

The argument about a 'technocratic paradigm', which 'allies the economy with technology and privileges the criterion of efficiency'(Pope Francis, 2023), was taken up again in the pope's penultimate message to celebrate the World Day of Peace, although the critique seemed not particularly pronounced. With regard to autonomous weapons systems, Francis expressed great concern, but in contrast to the ban argument he (Pope Francis, 2024) would make the following year, did not directly call for a ban, instead stating:

> Research on emerging technologies in the area of so-called Lethal Autonomous Weapon Systems, including the weaponization of artificial intelligence, is a cause for grave ethical concern. Autonomous weapon systems can never be morally responsible subjects. The unique human capacity for moral judgment and ethical decision-making is more than a complex collection of algorithms, and that capacity cannot be reduced to programming a machine, which as 'intelligent'as it may be, remains a machine. For this reason, it is imperative to ensure adequate, meaningful and consistent human oversight of weapon systems.
> (Pope Francis, 2023)

One can assume – and the heading of the section 'Shall we turn swords into ploughshares?' suggests this – that Pope Francis wanted to avoid distinguishing between 'better' and 'worse' weapons, so to speak. For him, all weapons seemingly were an evil. To a certain extent, the 'Campaign to Stop Killer Robots,' one of the leading NGOs in favour of a ban on AWS, can see itself supported by Francis's remarks, as it too has been calling for 'meaningful human control' in autonomous weapon systems for some time (see Amoroso & Tamburrini, 2020). That said, however, the challenge of the technocratic paradigm also lies in the fact that humans may lose their ability to exercise such control entirely (Koch, 2022). After all, even the controller needs criteria to judge whether a machine is behaving correctly or incorrectly. But where do these criteria come from? Possibly they come from machines again, and possibly even from the machines over which humans are supposed to exercise control. In such cases, it might be appropriate from an ethical perspective to remove humans completely from the chain of causation, because it is unreasonable to burden someone with (partial) responsibility for a cause-and-effect chain – a responsibility that they can never really fulfil. Or to put it more loosely: If there is no personal decision-making space left, it would probably be right not to pretend to be taking a personal decision either.

Francis's World Day of Peace message also remained vague with regard to the implications for peace policy. Although the danger of social friction through social media was alluded to, the risk of an international arms race of AI-supported weapons systems was not engaged directly (see Altmann & Sauer, 2017). While Francis was a long-time critic of the arms industry, his treatment of AWS alluded to a potential arms race implicitly only: 'In a word, the world has no need of new technologies that contribute to the unjust development of commerce and the weapons trade and consequently end up promoting the folly of war'(Pope Francis, 2023). However, there was one comment on the understanding of human dignity, which was important for the argument to be developed here: 'Fundamental respect for human dignity demands that we refuse to allow the uniqueness of the person to be identified with a set of data'(Pope Francis, 2023).

Our considerations above have thus shown that the question of human dignity does not primarily arise with regard to the persons attacked by autonomous weapons systems, but rather with regard to those who deploy and use such systems. However, the international campaign to ban such weapons systems has taken a different path. After initially focusing on the question of whether AWS would not accept too high a number of casualties, for example because they lack human empathy or are unable to distinguish between civilians and combatants (Human Rights Watch, 2012), the question of violation of dignity was also raised, particularly with regard to the people who are hit by the effects of such systems (see Rosert & Sauer, 2019). Of course, the previous objections could also be countered: On the one hand, AWS lack empathy, but on the other hand they do not act out of violent emotions such as anger or hatred. And the question of distinction between combatants and non-combatants also arises in humans, especially in the complex conflict situations of asymmetric wars. AWS should not be expected to act flawlessly, but only to act better than human soldiers when it comes to discernment. In connection with the distinction between civilians and combatants (or, more broadly, between legitimate and illegitimate targets) in armed conflicts, however, there is the question of the proportionality of collateral damage in attacks on military targets. This proportionality (non-excessiveness under international law) cannot simply be determined algorithmically (see Koch, 2019b).

Assessments respective trade-offs also presuppose value judgements: In order to weigh up two courses of action against each other, one needs to evaluate both. Such evaluations are not fixed once and for all in a 'vacuum,' but exist in a social space, are based on past judgements and cannot simply be made by computing power. It also should be noted that this problem is also encountered in AI-supported jurisprudence. Therefore, it is not a completely misguided question to ask whether civilians who are endangered, harmed or killed by military operations do not have some kind of claim that the judgement that led to them being endangered was made by a human being. But on the other hand, the non-fulfilment of a claim does not necessarily mean that the human dignity of the claimant has been violated. It also depends on the nature of the claim.

In the statement by Archbishop Jurkovič quoted above (Jurkovič, 2018), however, there is an indication of a claim whose violation actually concerns human dignity: the right to be recognized as a human being. This is a problem that particularly affects so-called 'social robots'nsuch as care robots or sex robots. The use of such devices may well be functionally successful: Dementia patients, for example, have been found to relax when in contact with the dummy seal 'Paro'(see Tergesen & Inada, 2010). In such limited fields of application, the argument that success can be achieved with AI-supported robotics may be ethically sound. However, such care robots only simulate care, but do not provide it. In order to be able to provide real care, they would have to have an idea of what it means to be a person suffering from dementia. Since they lack this idea, they cannot recognise the patient as a human being. Any object that is presented to them is just that: an object. Once again, in a scenario limited in time and space, this does not have to be an argument that beats all other arguments to the effect that the use would be unauthorised. But it would certainly be unauthorised to expose a human being to an environment that consists entirely of machine technology and at best simulates affection or recognition. Where AI comes into play as final or total, there is indeed a violation of human dignity.

It can be argued that this is precisely the case with AWS. Killing a human being is not an act limited in space and time, but for the victim the cancellation of space and time. It is therefore not unreasonable to see a formal minimum requirement in recognition for every act of killing, in addition to material legitimisation conditions (if there are any: just war theorists usually assume this; many pacifists deny this). This recognition is possibly 'only' the recognition of the other as an opponent or enemy, but it is not making the other invisible or negating him. For a fully autonomous armed drone, every target is just an object; metaphorically speaking, it makes no difference whether it is a sack of potatoes or a human being that the technical artefact is acting towards. 'For a machine, a human person, just like everything else, is merely an interchangeable set of data among others'(Jurkovič, 2018). According to the sentence quoted above from Francis's 2024 World Day of Peace Message, it is precisely in this view of a human being as a set of data points that dignity is violated. The total 'act' of the machine therefore not only takes the life of the person killed, but also violates her dignity.

In a well-known essay, Thomas Nagel (1974, p. 13) argued that violence must also be directed against an opponent as a person. The principle is 'that hostile treatment of any person must be justified in terms of something about that person which makes the treatment appropriate. Hostility is a personal relation, and it must be suited to its target.' This condition is meaningless from a functional point of view. From an ethical point of view, it may even seem completely wrong to those who favour consequentialism, such as utilitarians. For Nagel, it belongs to the field of 'absolutism which co-constitutes human morality from the subjective perspective (first-person perspective). We are not prepared to view our actions only in functional terms. It is not only important what consequences of our actions we 'produce,' but also that it is *us* (it is *me*) who acts in this or that way. There are therefore two competing perspectives here: The subjective one, which resists viewing people as material or data points and thereby reifying itself, and the objective one, which insists on this very reification, and which compares world states and demands that the always better world states be realised. Utilitarianism is paradigmatic for the objectivist perspective: only the external consequences of actions matter, not intentions. Ethical feelings only play a role insofar as they themselves are understood like external world conditions. This second perspective rejoices when evildoers have been rendered harmless – regardless of the means used. The first perspective demands that people (persons) deal with people (persons) as people (persons).

It is certainly questionable whether such theological-philosophical considerations are not too complex and too vague for political debate. In addition, the argument that human dignity has been violated also has a certain cultural relativity attached to it, which is already rooted in the different interpretations of dignity itself (see Sharkey, 2019). For political purposes – including purposes such as a ban on autonomous weapons systems on the basis of an agreement under international law – it is probably more expedient to point to the enormous security policy risks that this type of weapon brings into the world (see Dahlmann & Dickow, 2019). However, the task of religious authorities in dealing with pressing contemporary issues is probably not just to provide politically 'useful' arguments. It is therefore right to take up the anthropological self-image that is repeatedly addressed in questions of technology utilisation and to introduce in-depth criticism from there. The concept of dignity, which cannot be introduced in a dogmatically rigid way, can be a good point of reference here.

CONCLUSION

There can be no question that the ethics of AWS has turned into one of the most prominent themes of contemporary just war thinking. A diverse set of scholars has been arguing about the rights and wrongs of AI-enabled killing. Moreover, as recent reports from the ongoing war against Ukraine suggest, the use of AWS is no longer a hypothetical scenario. In this chapter, we have made an attempt to bring to bear the Catholic tradition of thinking about war and peace on the advent of AWS via the example of armed drones. Major voices in the Church have argued about the ethics of remotely-piloted drones, and there is an emerging body of remarks that grapples with the ethics of fully-autonomous weapons systems. Concentrating on the human dignity argument that lies at the very heart of Catholic social thought and that has repeatedly been invoked by Church leaders against AWS, we have contributed a theological-philosophical engagement with AWS that goes beyond the short statements provided by major voices in the Church to date. In so doing, we hope to have made a modest contribution toward an emerging Catholic perspective on AI-enabled weapons systems.

REFERENCES

Altmann, J., & Sauer, F. (2017). Autonomous weapons systems and strategic stability. *Survival. Global Politics and Strategy*, *59*(5), 117–142.

Amoroso, D., & Tamburrini, G. (2020). Autonomous weapons systems and meaningful human control: Ethical and legal issues. *Current Robotics Reports*, *1*, 187–194.

Baker, D. (2022). *Should we ban killer robots?* Polity Press.

Bode, I., & Huells, H. (2022). *Autonomous weapons systems and international norms*. McGill-Queen's University Press.

Bode, I., & Watts, T. (2023). *Loitering munitions and unpredictability: Autonomy in weapon systems and challenges to human control*. https://usercontent.one/wp/www.autonorms.eu/wp-content/uploads/2023/06/Loitering-Munitions-Unpredictability-WEB.pdf?media=1629963761

Braun, C. N. (2020). The Catholic presumption against war revisited. *International Relations*, *34*(4), 583–602.

Braun, C. N. (2021a). *Jus ad vim* and drone warfare: A classical just war perspective. In C. Enemark (Ed.), *Ethics of drone strikes: Restraining remote-control killing* (pp. 31–49). Edinburgh University Press.

Braun, C. N. (2021b). LAWS and the *mala in se* argument. *Peace Review: A Journal of Social Justice*, *33*(2), 237–246.

Braun, C. N. (2024a). Francis and the bomb: On the immorality of nuclear weapons. *Journal of Military Ethics*, online first, *23*(1), 1–10.

Braun, C. N. (2024b). Reading the 'signs of the time': Just war statecraft and the 'immorality' of nuclear weapons. *Studies in Christian Ethics*, online first, 1–12.

Brunstetter, D., & Braun, M. (2011). The implications of drones on the just war tradition. *Ethics & International Affairs*, *25*(3), 337–358.

Buhta, N., Beck, S., Geiß, R., Liu H-Y, & Kress, C., Eds. (2016). *Autonomous weapons systems: Law, ethics, policy*. Cambridge University Press.

Canadian Catholic Organization for Development and Peace. (n.d.). *10 principles of catholic social teaching*. https://stmikes.utoronto.ca/wp-content/uploads/2020/07/180-Catholic-Teaching-v2.pdf

Christiansen, D., & Sargent, C., Eds. (2023). *Forbidden: Receiving Pope Francis's condemnation of nuclear weapons*. Georgetown University Press.

Claudi, J. (2015, March 16). Künstliche Intelligenz beim Militär. Deutsche Welle. https://www.dw.com/de/k%C3%BCnstliche-intelligenz-beim-milit%C3%A4r/a-18288938

Dahlmann, A., & Dickow, M. (2019). Preventive regulations of autonomous weapons systems: Need for action by Germany at various levels. SWP Research Paper 3. https://www.swp-berlin.org/publications/products/research_papers/2019RP03_dnn_dkw.pdf

Davies, H., McKernan, B., & Sabbagh, D. (2023, December 1). Israel. 'The Gospel.' How Israel uses AI to select bombing targets in Gaza. *The Guardian*. https://www.theguardian.com/world/2023/dec/01/the-gospel-how-israel-uses-ai-to-select-bombing-targets

Deutsche Kommission *Justitia et Pax*, & Katholische Militärseelsorge. (2013, February 5). *Gemeinsame Erklärung*. https://www.bistum-essen.de/fileadmin/bereiche/za-kom/Gemeinsame_Erklaerung_zur_geplanten_Anschaffung_bewaffneter_Drohnen.pdf

Enemark, C. (2023). *Moralities of drone violence*. Edinburgh University Press.

Galliott, J., MacIntosh, D., & Ohlin, J. D., Eds. (2021). *Lethal autonomous weapons: Re-examining the law and ethics of robotic warfare*. Oxford University Press.

German Ethics Council. (2023). *Humans and machines. Challenges of Artificial Intelligence*. https://www.ethikrat.org/fileadmin/Publikationen/Stellungnahmen/deutsch/stellungnahme-mensch-und-maschine.pdf

Grünwald, R., & Kehl, C. (2020). *Autonomous weapons systems*. TAB-Fokus. https://publikationen.bibliothek.kit.edu/1000127167

Guardini, R. (2024; orig. 1950). *The end of the modern world: With power and responsibility*. Gateway Editions.

Hambling, D. (2023, October 13). Ukrainian AI attack drones may be killing without human oversight. *New Scientist*. https://www.newscientist.com/article/2397389-ukrainian-ai-attack-drones-may-be-killing-without-human-oversight/

Heinrich Boell Foundation. (2018). *Autonomy in weapons systems: The military application of Artificial Intelligence as a litmus test for Germany's new foreign and security policy*. https://www.boell.de/sites/default/files/boell_autonomy-in-weapon-systems_v04_kommentierbar_1.pdf

Himes, K. R. (2016). *Drones and the ethics of targeted killing*. Rowman and Littlefield.

Hollenbach, D. (1982). Nuclear weapons and nuclear war: The shape of the catholic debate. *Theological Studies*, *43*(4), 577–605.

Hrynkow, C., & Power, M. (2019). Are the popes leaving behind just war and embracing just-peace? *Peace Review: A Journal of Social Justice*, *31*(2), 238–246.

Human Rights Watch. (2012). Losing humanity: The case against killer robots. https://www.hrw.org/sites/default/files/reports/arms1112_ForUpload.pdf

ICRC. (2021). ICRC position on autonomous weapons systems. https://www.icrc.org/en/document/icrc-position-autonomous-weapon-systems

Jurkovič, I. (2018, April 9). *Statement by H.E. Archbishop Ivan Jurkovič, permanent observer of the Holy See to the United Nations and other international organizations in Geneva at the 2018 Group of Governmental Experts on Lethal Autonomous Weapons Systems (LAWS)*. https://docs-library.unoda.org/Convention_on_Certain_Conventional_Weapons_-_Group_of_Governmental_Experts_(2018)/2018_LAWS6a_HolySee.pdf

Kaag, J., & Kreps, S. (2014). *Drone warfare*. Polity Press.

Koch, B. (2016). Moral integrity and remote-controlled killing: A missing perspective. In E. di Nucci & F. Santoni de Sio (Eds.), *Drones and responsibility. Legal, philosophical, and sociotechnical perspectives on remotely controlled weapons* (pp. 82–100). Routledge.

Koch, B. (2019a). Die ethische Debatte um den Einsatz von ferngesteuerten und autonomen Waffensystemen. In I.-J. Werkner & M. Hofheinz (Eds.), *Unbemannte Waffen und ihre ethische Legitimierung* (pp. 13–40). Springer VS.

Koch, B. (2019b). Is proportionality a matter of attitude? A short reflection about an aspect of virtue in armed conflict. In B. Koch (Ed.), *Chivalrous combatants? The meaning of military virtue past and present* (pp. 139–160). Nomos.

Koch, B. (2022). Technikethik. In A. Merkl & K. Schlögl-Flierl (Eds.), *Moraltheologie kompakt. Grundlagen und aktuelle Herausforderungen* (pp. 340–350). Verlag Friedrich Pustet.

Koch, B. (2023). Drohnen und autonome Waffensysteme. Militärtechnologischer Wandel und ethische Herausforderungen. *Amosinternational*, *17*(2), 20–27.

Koch, B. (2024). Human dignity and, autonomous robotics. *What is the problem? Ethics and Armed Forces. Controversies in Military Ethics and Security Policy*, *1*. https://www.ethikundmilitaer.de/fileadmin/ethics_and_armed_forces/Ethics-and-Armed_Forces-2024-1.pdf

Lushenko, P., Bose, S., & Maley, W. (2021). *Drones and global order: Implications of remote warfare for international society*. Routledge.

McKenna, J. C. (1960). Ethics and war: A catholic view. *American Political Science Review*, *54*(3), 647–658.

Nagel, T. (1974). War and massacre. In M. Cohen, T. Nagel, & T. Scanlon (Eds.), *War and Moral Responsibility* (pp. 3–24). Princeton University Press.

National Conference of Catholic Bishops. (1983). *The challenge of peace: God's promise and our response*. https://www.usccb.org/upload/challenge-peace-gods-promise-our-response-1983.pdf

O'Connell, M. E. (2023). Banning autonomous weapons: A legal and ethical mandate. *Ethics & International Affairs*, *37*(3), 287–298.

Pontifical Council for Justice and Peace. (2004, April 2). *Compendium of the social doctrine of the church.* https://www.vatican.va/roman_curia/pontifical_councils/justpeace/documents/rc_pc_justpeace_doc_20060526_compendio-dott-soc_en.html#PRESENTATION

Pope Francis. (2015). *Encyclical Letter Laudato Sí.* https://www.vatican.va/content/francesco/en/encyclicals/documents/papa-francesco_20150524_enciclica-laudato-si.html#_ftnref87

Pope Francis. (2019, November 24). *Meeting for peace: Address of the holy father.* https://www.vatican.va/content/francesco/en/messages/pont-messages/2019/documents/papa-francesco_20191124_messaggio-incontropace-hiroshima.html

Pope Francis. (2020, September 25). Video message of his holiness Pope Francis to the seventy-fifth meeting of the General Assembly of the United Nations. https://www.vatican.va/content/francesco/en/messages/pont-messages/2020/documents/papa-francesco_20200925_videomessaggio-onu.html#_ftn16

Pope Francis. (2022a, March 18). *Address of his holiness Pope Francis to participants in the international congress promoted by the pontifical foundation Gravissimum Educationis.* https://www.vatican.va/content/francesco/en/speeches/2022/march/documents/20220318-fondazione-gravissimum-educationis.html

Pope Francis. (2022b, September 15). *Press conference on the return flight to Rome.* https://www.vatican.va/content/francesco/en/speeches/2022/september/documents/20220915-kazakhstan-voloritorno.html

Pope Francis. (2023, December 8). *Message for the 57th World Day of Peace 1 January 2024. Artificial Intelligence and Peace.* https://www.vatican.va/content/francesco/en/messages/peace/documents/20231208-messaggio-57giornatamondiale-pace2024.html

Ratzinger, J. (2004). *Worthiness to receive holy communion: General principles.* https://www.catholicculture.org/culture/library/view.cfm?id=6041

Reichberg, G. M. (2002). Is there a presumption against war in Aquinas's ethics? *The Thomist, 66*(3), 337–367.

Reid, C. J. (Ed.). (1986). *Peace in a nuclear age: The bishops' pastoral letter in perspective.* The Catholic University of America Press.

Rosert, E., & Sauer, F. (2019). Prohibiting autonomous weapons: Put human dignity first. *Global Policy 10*(3), 370–375.

Sharkey, A. (2019). Autonomous weapons systems, killer robots, and human dignity. *Ethics and Information Technology 21*, 75–87.

Skerker, M., Purves, D., & Jenkins, R. (2020). Autonomous weapons systems and the moral equality of combatants. *Ethics and Information Technology*, *22*, 197–209.

Tergesen, A., & Inada, M. (2010, June 21). It's not a stuffed animal, it's a $6,000 medical device. *The Wall Street Journal.* https://www.wsj.com/articles/SB10001424052748704463504575301051844937276

United States Conference of Catholic Bishops. (2013, May 17). *Letter to administration/congress on drones & targeted killings.* https://www.usccb.org/issues-and-action/human-life-and-dignity/war-and-peace/arms-trade/upload/letter-to-administration-congress-on-drones-2013-05-17.pdf

United States Conference of Catholic Bishops. (2015, May 11). *Letter to National Security Advisor Rice on a policy framework for the use of armed drones.* https://www.usccb.org/issues-and-action/human-life-and-dignity/war-and-peace/arms-trade/upload/letter-to-nsa-rice-from-bishop-cantu-re-drones-policy-framework-2015-05-11.pdf

7 The Paradox of Distance

AI, Violence, and Embodiment On and Off the Battlefield

Noreen Herzfeld

As I write Russia and Ukraine enter a ninth year of warfare, twenty-four months of intense fighting. The war in Gaza begins its second year, with 85% of Gaza's citizens displaced and more than 23,000 dead. In each of these conflicts, computer technology plays an increasingly large role, not just in winning ground or advantage on the battlefield, but also in winning the hearts and minds of public opinion around the world. Drones, cruise missiles, and unmanned reconnaissance vehicles are playing a larger part than ever in these conflicts. In the meantime, information and disinformation, videos, and soldier's voices on social media sites such as Twitter/X, Instagram, and Facebook bring these conflicts into our pockets and bedrooms and call those of us far away from the battles themselves to take a public stand. Do these technologies promote violence or nonviolence? Do they strengthen human solidarity and dignity or fracture them? To answer these questions, I begin with the words of Palestinian peace activist Ail Abu Awwad: "[Nonviolence] is based on humanity and responsibility." At the core of nonviolence is the recognition of the humanity of one's opponent and taking one's rightful responsibility in conflict situations (Nonviolence in the Time of War 2024, 25).

In this chapter, I will consider how computer technologies, in the forms of lethal autonomous weapons (LAWS), and social media, influence our recognition and relationship to the other, not only for those who fight, but also for those who merely observe the conflict. There is a paradox raised by computer technologies in terms of their distancing effects. While LAWS distance the soldier both physically and mentally from his or her opponent, thus retaining agency while minimizing or removing responsibility, social media does the opposite, bringing conflicts from around the world into one's home, promoting a sense of responsibility but also polarization and revulsion. Both make recognition of our shared humanity more difficult because through them we fail to encounter the other as fully embodied beings.

 DOI: 10.1201/9781003440253-7

LETHAL AUTONOMOUS WEAPONS AND THE FINAL DISTANCING OF WARFARE

Computer assisted weapons of war exist along a scale of autonomy. At one end are guided missiles launched by a human operator who selects the target. Computer algorithms then assist in determining the missile's trajectory. Weapons with slightly more autonomy have human operators "on the loop" rather than "in the loop." These weapons may choose their own target, yet operate under human supervision, allowing an operator to halt or alter the weapon's engagement. Fully autonomous weapons 'once activated, can select and engage targets without further intervention by a human operator' (Congressional Research Service 2020). To be fully autonomous, a system must be sufficiently opaque, 'such that, even when it is functioning perfectly, there remains some uncertainty about which targets it will attack and why' (Sparrow 2007, 64). Any system that does not exhibit this opacity is no longer autonomous, but completely determined by its programmers. While most autonomous weapons systems in use today still have humans in or on the loop, fully autonomous systems relying strictly on AI algorithms exist or are under development in several countries, including the United States and China; it is only a matter of time before they will be deployed more widely in battle (Scharre 2023, 5–7).

The potential advantages of such weapons are many. In a simulation dogfight staged by the United States Defense Advanced Research Projects Agency (DARPA) in which an AI competed against an experienced F-16 pilot, the AI weapon racked up 15 kills against none for the human. It showed superior precision in managing the aircraft, targeting its weapons, and was able to execute manoeuvres that would be impossible for a human to withstand (Scharre 2023, 2–3). Autonomous weapons also have the advantage of speed. AIs can process information faster than the human brain, and direct weaponry with a speed far beyond that of human muscles.

Consider a swarm of drones such as the Kargu-2, a small Turkish drone with image processing, machine learning, and the ability to carry an explosive payload. They can attack *en masse* and then rapidly disperse, leaving little target for a counterattack. The direction of the attack can be varied and loitering drones can bide their time or come and go in unpredictable ways. With image processing and machine learning, the attacking drones can observe and analyse the terrain, removing that advantage from the defender, and will be unaffected by fatigue, mental strain, or emotional compunctions.

But perhaps the greatest advantage of autonomous weapons is that they lower the need for human 'boots on the ground.' As an editorial in *The Economist* points out, 'a president who sends someone's son or daughter into battle has to justify it publicly, as does the congress responsible for appropriations and a declaration of war' (2010). If LAWS could make it possible to wage war without putting lives at risk is it ethically responsible to deploy human soldiers? Fully autonomous weapons represent a long progression in weaponry development that increasingly distances the soldier from his or her adversary, until it doesn't. Fighting with sword or bayonet required one to look the other in the eye. The crossbow and then the gun added a measure of distance, though one remained in the physical presence of the other. The advent of flight inaugurated a new era of warfare, releasing armies from a necessary physical presence on the battlefield. One need not see the enemy on whom one drops a bomb, though one still remains in the airspace above the target. Remotely operated drones add to this level of distance. A soldier can sit in a bunker in Nevada and unleash a missile or drone against an adversary who is half-way around the world.

AI is not the first technology to distance soldiers from their targets. But how this distancing occurs has psychological and moral implications, for distancing and dehumanization go hand in hand. Both the bomber pilot and the drone operator kill from a distance. Of the 90 airmen who flew on the atomic bombing missions of WWII, only one publicly declared remorse (Harrington 2020). On the other hand, in a 2020 study conducted by the US Air Force, 20% of drone operators reported emotional stress or PTSD (Philipps 2022). Why the difference? The drone operator can see the face of the 'Other'. Neal Scheuneman, a drone sensor operator who retired as a master sergeant from the Air Force in 2019 sums it up:

> A fighter jet might see a target for 20 minutes. We had to watch a target for days, weeks and even months. We saw him play with his kids. We saw him interact with his family. We watched his whole life unfold. You are remote but also very much connected. Then one day, when all parameters are met, you kill him. Then you watch the death. You see the remorse and the burial. People often think that this job is going to be like a video game, and I have to warn them, there is no reset button.
>
> (Philipps 2022)

The drone operator, while physically distant, remains mentally on the field of battle. He or she sees the face of the adversary. Philosopher and theologian Emmanual Lévinas underlines this importance of face-to-face encounter in our postmodern world: 'The relation to the face is straight-away ethical. The face is what one cannot kill' (Lévinas 1969). A face makes a person real and immediate. The drone operator retains the ability to see the face, hence the humanity, of the other. Fully autonomous weapons, in releasing soldiers from the mental decisions of the battlefield, release them from ethical obligations as well. It is for this reason that former USAF Major General Robert Latiff has called this transformation the crossing of a 'moral Rubicon' (Latiff & McCloskey 2013).

Georgia Tech roboticist Ron Arkin has argued that this is a Rubicon we must cross. Arkin believes LAWS have the potential to act more virtuously than humans. He cites a report from the Surgeon General's Office assessing the battlefield ethics of US soldiers and marines in which ten percent reported mistreating noncombatants and roughly thirty percent reported facing ethical situations in which they did not know how to respond (n.d.). Soldiers, under pressure, react emotionally: 'Fear and hysteria are always latent in combat, often real, and they press us toward fearful measures and criminal behavior' (Walzer 2015, 251). Lacking emotions, LAWS will act strictly on their programming, rather than out of panic or vengeance. They will follow orders more exactly. Image processing and machine learning would allow them to integrate information regarding a changing battle scenario faster than a human operator, thus allowing them to target the enemy with greater precision and fewer mistakes. This, Arkin believes, would result in fewer civilian casualties and war crimes, making the waging of war a more ethical endeavour and saving lives (Arkin n.d.).

Former lieutenant colonel David Grossman disagrees. Grossman notes that killing has, in the past, not come easily to human soldiers. He cites a study conducted by the US Army that found that only 15 to 20 percent of soldiers fired their weapons in combat in World War II. Fewer fired to kill (Marshall 2000). While these percentages have risen in subsequent wars, Grossman believes that failure to see the humanity of the 'Other' and physical distance from the 'Other' are the two factors that overcome our natural resistance to killing. Fully automated warfare that is 'anonymously fought by people who have never seen their enemy alive or dead and who will never know whom they have killed' exhibits both of Grossman's factors (Morganthau 2006, 250). Distance and dehumanization go hand in hand.

Pope Francis calls our attention to this coupling of distancing and dehumanization, describing our society as having a 'culture of indifference.' To follow Jesus, he writes that we must 'work and ask for the grace to build a culture of encounter, … that returns to each person their dignity as children of God, the dignity of living' (2016).

SEPARATING AGENCY FROM RESPONSIBILITY ON THE BATTLEFIELD

Action at a physical and mental distance presents a second moral issues in that it displaces human agency and obscures responsibility. Speaking as Vatican observer to the UN in Geneva, Archbishop Ivan Jurkovic stated that autonomous weapons present "the capacity of altering irreversibly the nature of warfare, becoming more detached from human agency, putting in question the humanity of our societies' (Catholic News Service 2018). Autonomous weapons not only distance combatants from one another. They also distance the act of killing from responsibility for that killing. If our weapons take over our agency in battle, can they also be said to take over our moral responsibility? For agency never stands alone. As philosopher Robert Sparrow notes, 'it is a necessary condition for fighting a just war, under the principle of jus in bellum, that someone can be justly held responsible for deaths that occur in the course of the war' (2007). While it is easy to ascribe the agency of killing to an autonomous weapon, it is much harder to determine where the moral responsibility might lie.

Despite Arkin's rosy predictions, computer programs are not foolproof. Eventually, an autonomous weapon will be involved in an accident or atrocity that seriously violates international law or crosses ethical boundaries. Recall that one of the stipulations for a weapon to be considered autonomous is that its choices and decisions must carry a certain degree of unpredictability. While the actions of an autonomous weapon may be foreseeable in most circumstances, they will not always perform as expected. Indeed, systems that depend on machine learning can be 'opaque even to the system's designers' (Crootof 2022, 1373). While they carry a certain degree of responsibility for creating such a machine, they cannot be held responsible for an action that they could in no way foresee (Herzfeld 2022, 80).

So who is responsible? If not the programmer, can the machine itself be held responsible? According to University of Richmond law professor Rebecca Crootof, 'traditional justifications for individual liability in criminal law—deterrence, retribution, restoration, incapacitation, and rehabilitation—do not map well from human beings to robots' (Crootof 2022, 1377). To call something a war crime it must have been 'wilfully' committed. Without consciousness, a machine cannot behave with intention, for from where would that intention arise? lacking emotion, the weapon itself has neither desires, needs, nor wants. Nor can it be punished since it feels neither mental nor physical pain. While it can certainly be decommissioned, it cannot feel remorse or any sense of responsibility. The age-old code of warfare coupled responsibility with risk. As T. R. Fehrenbach wrote, ‘a soldier’s destiny—which few escape—is to suffer, and if need be, to die’ (Quoted in Singer 2009, 332). A human who goes to war learns courage, the ability ‘to hold their ground when every instinct calls upon them to run away.’ There is a recognition of mortality, both in oneself, one’s comrades, and in the enemy. Responsibility in warfare relies on this recognition.

A machine could only be held responsible if it were conscious, and therefore intentional, or, lacking that, at least seemingly indistinguishable from humans in its decision making capacity. David Gunkel notes that a number of scholars argue that ‘as robots and AI get increasingly capable, sentient, and maybe even conscious, we will need to consider their interests and well-being in a way that is no different from that enjoyed by other persons’ (2024) He suggests we may need to shift from seeing rights as having an ontological basis to a stance where it is ‘social performance that determines status and standing by way of communal participation and interaction’ (Gunkel 2024) However, this obviates the ‘boots on the ground’ rationale for using autonomous weapons. As Sparrow notes, if our weapons were to become conscious, and therefore suffer, or act as though they were conscious and so, by Gunkel’s standards receive a status equal to humans, then they ‘would have become our soldiers and we should be as morally concerned when our machines are destroyed—indeed killed—as we are when human soldiers die in war’ (Sparrow 2007, 73).

Perhaps responsibility for any acts committed by an autonomous weapon must lie with the officer or commander who orders the deployment of such a weapon. Under current military law, commanding officers can be held indirectly responsible for the actions of those their command if they could have foreseen or prevented such actions. LAWS’ inherent unpredictability challenges this. Crootof asks: ‘Would fully autonomous weapons be predictable enough to provide commanders with the requisite notice of potential risk? Would liability depend on a particular commander’s individual understanding of the complexities of programming and autonomy?’ (2022, 1381). Here we have the crux of the matter. The deployment of autonomous weaponry severs the age-old connection between agency and responsibility. Since neither the programmer, the program itself, nor the officer who launches it can be reasonably held responsible for the actions of a fully autonomous weapon, seemingly no one can.

In a recent World Peace Day address, Pope Francis acknowledged this problem:

> Autonomous weapon systems can never be morally responsible subjects. The unique human capacity for moral judgment and ethical decision-making is more than a complex collection of algorithms, and that capacity cannot be reduced to programming a machine, which as ‘intelligent’ as it may be, remains a machine.
>
> (2024)

He suggests that the only recourse is to keep humans 'in the loop,' exercising some measure of ultimate control. Sparrow agrees, arguing that it would be unjust to deploy a weapon for which agency and accountability are separate (2007, 74–75).

INFLUENCING HEARTS AND MINDS OFF THE BATTLEFIELD

Now we turn to the other side of the computing paradox. If LAWS dehumanize the battlefield and carry agency while seeming to remove responsibility, the computers in our homes and pockets do the reverse. By bringing the faces and voices of the warzone onto our screens they seem to close the distance between us and the war zone. Our social media feeds teem with videos shot in the trenches of the Donbas and show us the faces of children starving under blockade in Gaza. Battles are no longer waged only on the battlefield; there is also a concomitant battle for our hearts and minds waged in cyberspace. While information warfare has long been a part of every battle, never before has it touched almost every citizen. And while it would seem that having instant information draws us closer to our fellow humans, too often it has the opposite effect, dividing those with only partial information, or disinformation into opposing camps.

As with autonomous weapons, information warfare has been part of a long technological development. Radio brought the voices of soldiers and the sounds of detonating explosions. Television brought pictures in real time. But the smartphone has carried this to a new level on two fronts. While television and radio relied on journalists with their access to heavy equipment and captured our attention primarily during the six o'clock or ten o'clock news, smartphones have enabled soldiers themselves to film attacks, conditions in war zones, and war's devastation in real time and promulgate these instantly to our social media feeds. Never has the face of the 'Other' felt so omnipresent.

This brings a sense of responsibility to those physically distant from the battle itself. Marshal McLuhan, commenting in the age of television, already noted that ' [t] oday, after more than a century of electric technology, we have extended our central nervous system itself in a global embrace, abolishing both space and time. ... This has heightened human awareness of responsibility to an intense degree.' He notes of those affected by marginalization or the privation of war that ' [t]hey are now *involved* in our lives, as we in theirs, thanks to the electric media' (1964, 5–7).

This sense of responsibility points to a second front, that of information warfare. And AI plays a strong role in this battle for hearts and minds. The advent of generative AI and Large Language Models (LLMs) has expanded the reach and availability of information and of disinformation. With a prompt of a single sentence ChatGPT will generate a very convincing paragraph or two on most topics, with no regard for truth or falsity. This has lowered the barrier for entry for those who wish to create disinformation. A state actor like Russia no longer needs to finance a stable of young tech-savvy professionals such as St. Petersburg's notorious Internet Research Agency. A single person can now quickly generate and disseminate fake news tailored to a variety of recipients.

Generative AI programs have also made significant progress in the production of deep fakes, video and/or audio clips that portray a person saying or doing things they never did. These have progressed far beyond early attempts that merely pasted one face on the body of another. Somewhat convincing fakes have been widely used by war propagandists on both sides in Ukraine. For example, Russia disseminated a fake video in March of 2022 of Ukrainian president Volodymyr Zelensky supposedly calling on his soldiers to lay down their arms. Meanwhile, Ukrainian sources put out a steady stream of supposedly heroic feats, including using footage from a video game to support stories of the 'Ghost of Kiev,' a pilot who supposedly shot down six Russian aircraft (Scharre 2023, 129). While these fakes did not fool experts, who quickly unveiled them, they were not so easily dismissed by the casual observer. In the words of GPT3 itself, 'AI generated content will continue to become more sophisticated, and it will be increasingly difficult to differentiate it from the content that is created by humans.'

This has led to a situation where it can be hard to distinguish truth from falsity, where 'truth is whatever the powerful say it is' (Scharre 2023, 118–119). While generative AI programs present fiction as fact, propogandists use public confusion over what is real to also label fact as fiction. The Russian Ministry of Foreign Affairs claimed that the extensive video evidence of the massacre in Bucha was staged by crisis actors and 'another #hoax by the Kiev regime for the Western media' (Porter 2022). Similarly, numerous fake claims regarding the war in Gaza have been unintentionally shared by journalists, officials and concerned citizens, including footage from previous conflicts, incorrect geolocations and edited content.

Despite these incidences, these two conflicts have been documented in real time by soldiers and civilians in a way we have never seen before, bringing both triumphs and anguish to a worldwide audience. This has, per McLuhan, engendered a sense of responsibility in many who are far from the field of conflict. Unfortunately, the emotional nature of social media and its polarized dissemination has often led, not to greater understanding, but to a hardening of attitudes toward the opposing side. Dutch scholars Mykola Makhortykh and Maryna Sydorova documented social media usage in pro-Ukrainian and pro-Russian online communities during the 2014 conflict in Eastern Ukraine. They found that while the Ukrainians interpreted the violence as a limited military action against insurgents, the pro-Russian community saw it as a war against Russian speaking citizens. This propagation of differing explanations played a role, in their opinion, in forming the divergent views in Ukraine and Russia that led to the current war. (2017, 359–381). Similarly, the widespread dissemination of footage from the October 7 attack on Israel by Hamas, as well as footage of dying Palestinian children and physical devastation in northern Gaza has led all too often not to simple compassion for those Israelis and those Palestinians who have suffered, but to revulsion and demonization of one side or the other.

In *Totality and Infinity*, Levinas writes that rhetoric, in so far as it is self-centred and bent on persuasion, fails to truly see the Other and is thus as much an instrument of violence as other weapons (1969, 70). Speech that arises from ideological predispositions and distortions 'masks' the face of the Other. Thus, while social media may allow us to 'see' the Other as never before, it all too often is used to deny the alterity of the Other, thus denying their humanity. In the assumption that we take responsibility by weighing into a conflict that is not our own, we might instead by merely fanning the flames, adding verbal violence to that which is physical.

CONCLUSION: WE ARE EMBODIED CREATURES

In the encyclical *Pacem in Terris*, Pope John XXIII writes:

> Everyone, however, must realize that, unless this process of disarmament be thoroughgoing and complete, and reach men's very souls, it is impossible to stop the arms race, … But this requires that the fundamental principles upon which peace is based in today's world be replaced by an altogether different one, namely, the realization that true and lasting peace among nations cannot consist in the possession of an equal supply of armaments but only in mutual trust.
>
> (1963)

Mutual trust arises when we fully acknowledge the humanity of the Other, a mutuality that surpasses the alterity of the other. Once we admit the Other as equally human we are called into a relationship of responsibility, one, as Levinas notes, whose bedrock rests on the commandment 'thou shalt not kill' (Murray 2003, 4). Pope Francis calls us to a 'culture of encounter' in which we not only see the face of the Other, but also experience the fullness of their embodied humanity: 'If I do not look, — seeing is not enough, no: look — if I do not stop, if I do not look, if I do not touch, if I do not speak, I cannot create an encounter and I cannot help to create a culture of encounter' (2016). Autonomous weapons and social media both perpetuate violence because they fail to address at the deepest level that we are embodied creatures. Whether AI distances us from encountering the face of the other with autonomous weapons or seems to bring the face or voice of the other near through social media, both fall short of a fully authentic encounter.

What makes an encounter fully authentic? Swiss theologian Karl Barth outlined four criteria, namely, to look the other in the eye, to speak to and hear the other, to aid the other, and to do it gladly. For Barth, the first criterion, to look the other in the eye, is not just one of many but the bedrock on which all the others rest (1958, 250–271). We cannot fully listen, aid, or empathize unless we encounter the other and admit to ourselves as embodied beings. MIT sociologist Sherry Turkel agrees. She writes, 'Face-to-face conversation is the most human—and humanizing—thing we do. Fully present to one another we learn to listen. It's where we develop the capacity for empathy. It's where we experience the joy of being heard, of being understood' (2015, 3). It is only face-to-face with the other that we fully apprehend both the humanity and the alterity of the other and come to recognize our responsibility to them. Levinas writes, 'The Other becomes my neighbor precisely through the way the face calls to me, summons me, begs for me, and in so doing recalls my responsibility, and calls me into question' (1989, 83).

Ultimately, that responsibility is not to be found in words but in deeds, deeds that go far beyond 'thou shalt not kill.' We are called, in the words of Jesus, to feed the hungry, clothe the poor, heal the sick, and be present with those who suffer (Matthew 25:35–40). These are highly embodied acts, acts that demand our compassion, but also our resources. To the extent that new weapons or social medias drain our coffers and our time, they represent a hidden violence. As American President Eisenhower said,

> Every gun that is made, every warship launched, every rocket fired signifies in the final sense a theft from those who hunger and are not fed, those who are cold and are not clothed. The world in arms is not spending money alone. It is spending the sweat of its laborers, the genius of its scientists and the hopes of its children.
>
> (Quoted in Singer 2009, 436)

REFERENCES

Arkin, R. (n.d.). *Governing Lethal Behavior: Embedding Ethics in a Hybrid Deliberative/Reactive Robot*. Georgia Institute of Technology, College of Computing. Technical Report GIT-GVU-07-11. Retrieved from www.cc.gatech.edu/ai/robot-lab/online-publications/formalizationv35.pdf

Barth, K. (1958). *Church Dogmatics: The Doctrine of Creation Part 2* (Vol. 3). (G. Bromiley, T. Torrance, Eds., J. Edwards, O. Bussey, & H. Knight, Trans.) Edinburgh: T&T Clark.

Catholic News Service. (2018, November 28). Vatican official: Prohibit 'killer robots' now before they become reality. *The Tablet*. Retrieved from www.thetablet.co.uk/news/11072/vatican-official-prohibit-killer-robots-now-before-they-become-reality

Church of the Brethren Ministry and Mission Board. (2013, March). *Resolution Against Drone Warfare*. Retrieved from www.brethren.org/about/statements/2013-resolution-against-drones.pdf

Congressional Research Service. (2020, December 1). Defense Primer: U.S. Policy on Lethal Autunomous Weapon Systems. *Focus*. Retrieved from https://sgp.fas.org/crs/natsec/IF11150.pdf

Crootof, R. (2022, October). War Torts. *New York University Law Review*, *97*(4), 1063.

Drones and Democracy. (2010, October 1). *The Economist*. Retrieved from www.economist.com/babbage/2010/10/01/drones-and-democracy

Gunkel, D. (2024). Should Robots Have Standing? From Robot Rights to Robot Rites. *Frontiers of Aritificial Intelligence and Applications*.

Harrington, A. (2020, August 6). The Hiroshima Pilot Who Became a Symbol of Antinuclear Protest. *New York Times*. Retrieved from https://www.nytimes.com/2020/o8/06/magazine/hiroshima-claude-eatherly-antinuclear.html

Herzfeld, N. (2022). Can Lethal Autonomous Weapons Be Just? *Journal of Moral Theology*, *11*(Special Issue 1).

Latiff, R., & McCloskey, P. (2013, March 14). With Drone Warfare, America Approaches the Robo-Rubicon. *Wall Street Journal*. Retrieved from www.wsj.com/news/articles/SB10001424127887322412850457834633324614559O

Lévinas, E. (1969). *Totality and Infinity*. Pittsburgh: Duquesne University Press.

Levinas, E. (1989). Ethics as First Philosophy. In S. Hand (Ed.), *The Levinas Reader* (S. Hand, & M. Temple, Trans.). Oxford: Blackwell.

Makhortykh, M., & Sydorova, M. (2017). Social Media and Visual Framing of the Conflict in Eastern Ukraine. *Media, War & Conflict*, *10*(3). doi:10.1177/1750635217702539

Marshall, S. (2000). *Men Against Fire: The Problem of Battle Command*. Norman, OK: University of Oklahoma Press.

McLuhan, M. (1964). *Understanding Media: The Extensions of Man*. New York: McGraw-Hill.

Morganthau, H. (2006). *Politics among Nations*. New York: McGraw-Hill.

Murray, J. (2003). *Face to Face in Dialogue: Emmanuel Levinas and (the) Communication (of) Ethics*. Lanham, MD: University Press of America.

Nonviolence in a Time of War. (2024, January). *Sojourners*, p. 25.
Philipps, D. (2022, April 15). The Unseen Scars of Those Who Kill Via Remote Control. *New York Times*. Retrieved from https://www.nytimes.com/2022/04/15/us/drones-airstrikes-ptsd.html
Pope Francis. (2016, September 23). For a culture of encounter. *L'Osservatore Romano*(38). Retrieved from https://www.vatican.va/content.francesco/en/cotidie/2016/documents/papa-francesco-cotidie_20160913_for-a-culture-of-encounter.html
Pope Francis. (2024, January 1). *MEssage of His Holiness Pope Francis for the 57th World Day of Peace*. Retrieved from https://www.vatican.va/content/francesco/en/messages/peace/documents/20231208-messaggio-57giornatamondiale-peace2024.html
Porter, T. (2022, April 4). Russia cclaims it did not massacre civilians in Ukraine, citing conspiracy theories that evidence wars manipulated or filmed with crisis actors. *Business Insider*. Retrieved from https://www.businessinsider.com/russia-cities-conspiracy-theories-to-denying-massacre-in-bucha-ukraine-2022-4
Scharre, P. (2023). *Four Battlegrounds: Power in the Age of Artificial Intelligence*. New York: Norton.
Singer, P. W. (2009). *Wired for War: The Robotics Revolution and Conflict in the 21st Century*. New York: Penguin.
Sparrow, R. (2007). Killer Robots. *Journal of Applied Philosophy*, *24*(1), 64.
Turkel, S. (2015). *Reclaiming Conversation: The Power of Talk in a Digital Age*. New York: Penguin.
Walzer, M. (2015). *Just and Unjust Wars*. New York: Basic Books.

8 Actions and Accidents in the Future Battlefield

The consequences of AI for Military Chaplaincy and Jus in Bello

Kenneth Wilkinson-Roberts

This chapter explores the challenges that Artificial Intelligence (AI) poses for Just War Theory (JWT) from the perspective of Military Chaplaincy. I focus on *jus in bello*, the section of JWT which assesses activity in war. This chapter argues that *jus in bello* is vulnerable to several of the key characteristics of AI, creating conditions for unjust activity in a variety of ways. To do this, I explore how the viewpoint of military chaplaincy helps us to identify these challenges. By utilising the Chaplain's role in mediating *jus in bello*, and their focus on moral and pastoral challenges, I identify two areas of concern for JWT: action and accidents. Similarly, I draw on the work of Paul Scharre to articulate and evaluate the nature of the battlefield in which AI will be deployed. In his recent book *Army of None* (2018), Scharre offers a comprehensive examination of the place of robotics, AI and other emerging technologies in future wars, which form the foundation of this chapter. By drawing together Scharre's insights on AI and the perspective of Military Chaplaincy, this chapter will expose a number of challenges to JWT, asking urgent questions about how Just Wars should be maintained in an AI-saturated future battlefield.

Briefly, Just War Theory is an approach to warfare and warfighting which sets out moral, legal, and ethical frameworks for the different stages of conflict at the state level. Broadly, it is a framework which aims to morally justify violent conflict and protect smaller – typically state – powers from being overwhelmed by the actions of more powerful states. Although these guidelines were once steeped in rich theological thinking originating from Thomas Aquinuas, contemporary JWT has become somewhat secularised within Security discourses, leaving the theory philosophically rich but religiously ambiguous (Payne, 2021; Coker, 2015). Of course the theory remains theological within religious, philosophical and historical discourses as displayed throughout this book. The theory has also been transposed into contemporary religious traditions, alongside modern articulations of distinctive traditions independent of their Christian name sake (including Islam, see Fyfe, 2017; Ahmad, 2006). The exact reasons why just actions in war are just in each of these different JW

 DOI: 10.1201/9781003440253-8

traditions, are not the concern of this chapter. As such, I am taking as given that wars should be just, and enacted according to the contemporary JWT. My concern, instead, relates to how well this can be achieved whilst utilising the power of artificial intelligence in current and future wars.

In particular, this chapter explores some of the challenges to JWT posed by AI, from the perspective of military chaplaincy. AI has been a major area of innovation in recent years, with the UK Armed Forces focusing heavily on AI research and deployment. Indeed, the 2021 Defence Command Paper discusses how AI will play a major role in the military of the future, stating that 'future conflicts may be won or lost on the speed and efficacy of the AI solutions employed' (MOD, 2021: 42.). AI is not only useful in a combat setting. In fact, most AI is not yet developed enough to operate fully autonomously on the battlefield. Instead, it is being used to streamline information processing and facilitate the operations of human warfighters and decision making. For, AI is not only a lethal, combative technology, but will be employed throughout every area of military activity; from combat to accounting. We cannot, therefore, only think of AI as an operational asset or direct means of lethality. This chapter will take a general approach to artificial intelligence, approaching this topic not from the focus of a specific AI system or platform, but from the viewpoint of specific human actors.

Before moving onto the rest of the chapter, we should pause for a note on positionality. Many of the distinctions made in this chapter exclude crucial voices and perspectives, as is inevitable with any single book chapter. Military chaplaincy and the JWT are diverse and complex areas of research, branching far beyond their western and Christian discoursal origins. This chapter will focus on general understandings of the JWT, military chaplaincy, war and AI, using Christian understandings of ministry and ethics, as well as western conceptions of the future of war. This is not from a desire to homogenize knowledge in this area, but rather to prioritise material regarding the challenges of AI, from my own perspective as a Christian practical theologian and security thinker. The role of Chaplains should not be limited to only the language and concerns of Christian practical theology. As this would miss the vital voices, ideas, and experiences of chaplains from other religious traditions. A focus on Christianity alone – whether explicit or implicit – in Chaplaincy research is artificial at best and biased at worst. As such, even studies which do focus on Christian chaplaincy must acknowledge this limitation. The same is also true for the study of Just War. As, even though Just War has its historical roots in the Catholic tradition, it has since been developed, secularised, and transposed into other religious traditions to theologically assess state level conflict outside of the Catholic and western traditions. This chapter will not provide the multiplicity and diversity that I have just advocated for. This is – for me – the most significant limitation of this chapter. I have deemed it necessary to concisely and coherently represent the voice of military chaplaincy regarding AI's implications for *jus in bello*. Likewise, I would rather signpost you to experts in different traditions of chaplaincy. Namely, Michalowski (2015), Lewis (2022), and Feuer (2022), who explore chaplaincy within the Islamic, Jewish and Buddhist traditions respectively. Read alongside this chapter, these contributions can provide a more accurate and diverse picture of chaplaincy's relationship with warfare, just war and AI than I can provide alone.

MILITARY CHAPLAINCY AND JUST WAR

Military Chaplaincy is hard to define. Every context, individual chaplain, and religious tradition presents differently. This leads to a flexible conception of who a military chaplain is and what they do. In an institutional sense, the military chaplain is a (typically ordained) civilian appointed from a specific religious tradition to minister to a specific group of personnel. They are located in the workplace of these military personnel and accompany them on their day-to-day activities, including on deployments oversees. Beyond this institutional definition chaplains are tasked with overseeing pastoral, spiritual and religious needs and care (Swift et al., 2015). Winfield describes American Military Chaplaincy as including:

> Leading worship services, liturgies, and rites; procuring religious items, equipment and food; advising the command on religious accommodations and religious conflicts; advocating for the promotion of moral and ethical behaviour within a unit; providing one-on-one spiritual counselling; training religious lay leaders to meet diverse needs.
>
> (Winfield, 2022: 34–5)

Here, the role of chaplains is articulated in terms of the activities that they do, in providing for specific moral, pastoral, religious and morale concerns in practical ways. These functional definitions are typical when defining military chaplaincy – as well as ministry in any context. However, this framework limits our understandings of chaplaincy to their practical outworking. Metaphysically, military chaplaincy is rooted in a theological framework, as well as a military one. Theologically, chaplaincy is based on the desire for those engaging in combat to be spiritually and morally cared for. In this way, JWT and military chaplaincy have similar normative roots: to theologically attend to the souls of military actors embarking on state sanctioned violence. JWT does this through a legal framework, whilst chaplaincy uses a ministerial one.

From a security perspective, Chaplains are responsible for ensuring readiness and resilience. They do so by mediating and resolving the moral, ethical, spiritual, and religious obstacles to warfighting through supporting personnel and ensuring that their religious needs are met. This runs parallel to the role of the medic who is responsible for attending to medical obstacles to operations. Indeed, Snape begins his history of British Military Chaplaincy by discussing the words of Anthony Farrer-Hockley in 1951, where if there:

> … was a question that soldiers were reluctant to ask their regimental officers because 'those concerned had no wish to give the impression that they were looking for an excuse to leave their comrades in the midst of operations in a hard climate'. However, they had fewer inhibitions when confiding in chaplains. According to Farrer-Hockley, 'Their relationship with the chaplains was different; in such a matter he was better informed as to their state of mind'.
>
> (Snape, 2008:1)

These chaplains became experts in the conditions of the soldiers around them, and so could ensure that they were able to carry out their duties without threatening disciplinary action or unit cohesion. In a similar vein, military chaplains are tasked with maintaining morale amongst military personnel. Morale is vital to the smooth running of military operations, whether they are day-to-day taskings or battlefield activities. Again, this role revolves both around a pastoral responsibility to care for personnel, and a military responsibility to ensure that these troops are able to do their jobs effectively. Here, we can see the mutually beneficial – though also potentially conflicting – nature of Chaplaincy in the armed forces. Moreover, we can begin to understand some of the ontological aspects of military chaplaincy. Chaplains are worship leaders, moral/ethical guides, and pastoral caregivers. They make themselves available, so to attend to the needs of the personnel around them, and they liaise between their unit, and its superiors. It is these roles and responsibilities (and many more) that inform the functioning of military chaplaincy, alongside the context in which they are working, and the specifics of the personnel they care for. For the purposes of this chapter, I will maintain this multilayered understanding of chaplaincy, but with a focus on their role as moral and ethical leaders; for this is the aspect that most relates to *jus in bello*.

Hitherto, *jus in bello* has only been explained in very simple terms. To expand a little, *jus in bello*, or *justice in war*, is the section of JWT which governs the activities of war. This aims to ensure that when war has to take place – according to *jus ad bello* – it does so in a just way. Of course, in a literal sense, this is impossible. The nature of war requires violence and destruction (Clausewitz, 2012: Chapter 1), which are usually seen as unjust in an abstract sense. What *jus in bello* ensures is not that nothing violent happens within war, but that when it does, it happens in a way which is proportional, discriminatory, and limited. These aims are often articulated at the strategic level, with compliance determined by the kinds of strategies that states employ. However, unlike *jus ad* and *post bello*, the warfighters involved in enacting strategy on the ground are also responsible for ensuring that their conduct is just. As such, it enters the domain of the chaplain as she ministers to the warfighters that must enact orders alongside the demands of behaving justly.

There are two major aspects of military chaplaincy that relate to *jus in bello*: (1) their role as moral leaders; and, (2) their role as moral monitors. Winfield describes these aspects of chaplaincy as being ‘a moral compass for the unit ensuring that service members adhere to the rules of engagement and answering larger ethical questions’ (Winfield: 35). Moral leadership is a vague role in chaplaincy, relating to role modelling as well as tasks like leading moral education. In the former case, Chaplains demonstrate through their behaviour how to act morally. Chaplains are also moral advocates for their personnel, demonstrating moral leadership through action on behalf of those that they minister to.

Related to this is designing and running education surrounding moral and ethical issues. This sometimes includes teaching military personnel about the ethical challenges and dilemmas involved in their work, as well as teaching personnel about moral decision making and judgement more generally, often in the form of one-to-one pastoral dialogue. These activities familiarise military personnel with the moral and ethical dimensions of their work, and empowers them to access complex operational situations, and make ethical choices. This role in education and leadership is not the sole remit of the chaplain. Regimental authorities, training institutions and external experts are also involved in moral and ethical education of personnel throughout their time in the military. The difference with the chaplain is often their bespoke, as Snape puts it, 'expertise and authority' in moral matters (Snape: 104), and their close relationship with personnel, which makes the chaplain's place in *jus in bello* so vital and distinctive.

ACTIONS

In many ways, *Jus in Bello* is concerned with actions. What actions do warfighters take in battle? How are these actions achieved? However, JWT is not always interested in what these actions are specifically. For instance, bombing urban targets in of itself is not forbidden. Rather, it is the intention, circumstances, and implications of actions that are important. If a target were on the outskirts of a city, in a conflict where bombing such a target was proportional to the threat, then the action may be justified. But, if the target was next to a school, then it would probably not be justified.

Artificial Intelligence is said to be of great benefit for *Jus in Bello*, primarily because it facilitates higher levels of accuracy and precision in targeting, and faster calculations of collateral damage. As such, activities involving AI are intended to be more appropriate when it comes to doing *just* actions in war. This is a beneficial idea in theory, but it relies on the ability of AI systems to do this well. For instance, facial recognition software is wrapped up in the biases of western culture, making accurate facial recognition questionable, particularly when assessing black and brown faces, women's faces, and the faces of those whose appearance does not match traditional social norms (Buolamwini, 2023; Beauchamp, 2019). As such, more accurate facial recognition is required before it can be claimed to be ethically beneficial for deployment in active conflict. Likewise, AI targeting creates problems strategically. Samuel Moyn writes:

> In spite of the possibility that machines will bring us even more humanity in war, the arguments of the new abolitionists are strong… [autonomous machines] will become even more grievously fatal than before, and to more innocent people. Either they will – like the airplane – kill on a more massive scale than before, or they will – like the drone – allow discrimination of targets in some episodes but rain down more death through more operations.
>
> (Moyn, 2022: 320)

Indeed, AI powered technologies promise otherwise unprecedented levels of speed, agility and cognitive power. Making them excellent investments and key resources in shaping and undertaking new strategy. However, Moyn articulates the double-edgedness of humaneness, where attempting to behave more justly might result in less morally justifiable action; despite remaining satisfactory in the just war framework.

Chaplains are inevitably concerned with these 'higher level' moral and ethical questions, especially those ministering to personnel at the strategic level. So far, much of the literature regarding AI and military chaplaincy has remained at this national and strategic level, discussing broad Just War topics and issues of proportionality and international legality (Fairclough, 2019). There is less written about the more immediate challenges AI can pose for the '*front-line*' military chaplain (operating at the tactical level of warfighting). In focusing on 'front-line' chaplaincy, I am not drawing any normative distinction between the chaplains in different institutional locations. Rather, this chapter focuses on one specific kind of instance where chaplains and personnel interact with AI. Revd David Fisher writes, soon after the deployment of drones in the UK Armed Forces, that drones pose a threat to *just* decision making, through tempting personnel into making immoral choices (Fisher, 2013). In particular, Fisher argues that, 'with the advent of drones and the so-called roboticization of war, warfare is now so different that the old concepts and catagories simply no longer apply… changed so radically that ethics just cannot keep up' (Fisher: 86). There are a range of reasons for this, including Moyn's concern that it will lead to more destruction. Though Fisher also argues that it is the warfighter's ability to make and enact ethical choices that are at risk. He argues that this is because of the gamification of war, due to videogame like dynamics of drone operations. This argument has since been debunked (Chamayou, 2015; Lee, 2019), but Fisher's anxiety surrounding a rupturing of moral decision-making stands because of the dynamics of drone operations. Specifically, Fisher is concerned with the metaphysical distance between the drone pilot and her target, leading to a disruption in traditional military narratives of heroism, bravery and courage. This – coupled with the limited risk of the drone pilot being harmed – can lead to reckless and poor moral decision making (Fisher: 86–88).

Fisher's thesis, alongside Mohl's, creates an interesting challenge for *jus in bello*. They highlight how autonomous technologies create a divergence in the moral characteristics of warfighting. At the strategic level, the dynamics of drones (and also autonomy) creates a more just kind of warfighting, with more precision and accuracy. However, at the tactical level, it can result in poor and/or reckless decision making, potentially leading to unjust warfighting, according to the JWT. Of course, war is always complex morally, and disconnects between strategy and tactics are common. However, what AI could create a situation where it is required at the strategic level, because of its benefits for targeting and cognitive pace. Whilst simultaneously leading the warfighters who enact this strategy to behave unjustly. As a result of this challenge, Fisher argues for an emphasis on moral courage. That is, courage to behave morally even when tempted or under-pressure to do otherwise (Fisher: 88). For Fisher, it is the military chaplain who must be an expert at this (to ensure their own moral decisions, and) so that they can mediate, monitor and teach moral courage to personnel. Therefore, the deployment of AI in future battlefields will require an intensifying of moral leadership from chaplains, so to avoid the moral ruptures that may emerge from the use of 'humane' technologies like AI.

At this point, Security scholars would argue that this is an unnecessary anxiety because, in the future battlefield, the AI itself will be making the crucial moral decisions (Coker, 2015; Scharre, 2018). Here, humans would be relieved of making and living with difficult moral choices, as these will be given over to the more cognitively advanced AI. Plus, they can do so without any possibility of psychological or pastoral damage (Scharre: 273–4). However, aside from the technological challenges of building such a sophisticated AI, and the complex moral framework that this AI would have to be programmed with, there are elements to these suggestions which ought to make military chaplains – and military decision makers – worry. For, it takes away from the profession of warfighting, outsourcing it to AI technologies at the cost of human warfighters. Making moral decisions is not a complication of warfighting. It is part of it, inextricably. Payne writes that, the responsibility of making morally challenging choices is part of what soldiers are trained for, it is part of what it means to be a soldier. Therefore, outsourcing this to a technological agent takes that away from the human warfighter (Payne, 2021: 219–256). Though this may be necessary to keep personnel safe, or to ensure that action keeps up with the escalating pace of war, it still affects the nature of warfighting in a fundamental way. Likewise, puncture the ontology of warfighting and endanger retention and resilience (the cornerstone of military chaplaincy from a strategic perspective).

Regarding *jus in bello* in this context, the relocation of moral responsibility from humans to AI creates challenges for the JWT, because it disrupts the ontological foundations of the just war tradition. According to Payne, the JWT has an 'inseparable human essence', with a morality and legality that 'originates in the human condition' (Payne, 2021: 232). This is inevitable when humans are making rules to govern the intensity and complexity of war. The problem is that when these rules and laws are programmed into AI, we ask AI to operate according to a framework that it cannot relate to. Thus, creating a technological agent to embody human laws and rules. For Payne, these laws and rules are not purely deontological. They require emotional and moral values to interpret them, such as moral courage, compassion, empathy and bravery. For instance, compassion and empathy can help soldiers determine whether to fire on a combatant out of combat, or whether to escalate a stand-off (Scharre: 207–8; Payne: 233). Without these moral instincts, AI may make choices which are morally just but strategically problematic, or which are overly aggressive at times when restraint is required. Of course, humans are also guilty of making moral mistakes: firing out of anger, misjudging a situation, hesitating because of cowardice. However, the relative power and pace of AI-based weapons may mean that the damage done by an AI's aggression are greater than that of one soldier's, thus again risking unjust action. The question here is whether the benefits of AI are worth the moral risk? For chaplains, this is a core question, as they will be the immediate port of call for personnel operating alongside AI and dealing with the consequences of their lack of empathy, or courage.

Zweibelson takes these concerns a step further, looking past the immediate future, to a time when strategic and tactical choices are made purely by AI, because of their superior cognitive abilities and speed (Zweibelson, 2023). Here, Zweibelson tracks the progression of military decision making from purely human determined, through a period of human and AI teaming, to a single AI overseeing other autonomous military agents. He calls this final stage the singleton:

> Bostrom explains that a singleton is an entity that becomes the single decision-making authority at the highest level of human organisation. This assumes the entirety of human civilization… a singleton is something that is able to take total control of human civilisation, or at least those that are reachable to be controlled, so that a world order is instilled by the design of the singleton and executed through complete rationalisation.
>
> (Zweibelson, 2023: 16–17)

For the purposes of this chapter, I am only interested in an AI singleton responsible for military and defence decision-making, not a total singleton like Bostrom defines. For Zweibelson, this singleton, '... is well beyond a game-changing "super weapon" or something that requires novel strategy in warfare. A singleton paradox transforms war toward something potentially unrecognizable or even comprehendible to ordinary humans' (Zweibelson, 2023: 29). Here there is a shift in ontological and strategic emphasis, from an AI making choices about a human conception of war, for human purposes, to one of technological conceptions and purposes. If we turn this towards JWT, Zweibelson's singleton intensifies Payne's arguments about the ontology of JWT. For, while AI is making decisions about human defined society and politics, just war tradition makes sense. However, once war has accelerated and technologized beyond the scope of humans, the JWT loses its relevance. This may not be a problem for the JWT more broadly, because much of the warfighting by this stage will be technologized anyway. Thus diminishing the need for, say, proportionality, if all the combatants are machines, and if no civilians are in range of the attack. In this context, the human suffering of war could be removed, thus removing the need for AI to behave ethically in the first place.

Zweibelson acknowledges that an AI singleton in security is theoretical, and a long way off technologically. However, imagining it offers us a glimpse into the condition of the JWT in the future. For, it shows us how precarious these ethical frameworks are as technology advances and saturates the battlefield. It is inevitable that human warfighters will be affected by this shift in military demographic, emphasis and ethics, resulting in Mohl's fears of unjust strategy, Fisher's predictions of reckless and immoral behaviour, and Zweibelson's warnings about a shifting metaphysics of war. These shifts will intensify the moral role of chaplains, in order to maintain just action and mediate these various moral dangers in the midst of fighting.

ACCIDENTS

However, the above discussion rests on an image of warfighting that is somewhat predictable and intentional. In this imagining of war, an action is decided and enacted, with specific implications which we can predict and measure, to make morally just and strategically sound decisions. However, this is not a true depiction of war, especially not in the future. Firstly, my discussion of actions assumes that the kind of technologies I am concerned about can be built in the first place, and if so, that they work. The former of these is a question for engineering, but the latter ought to concern theological and security studies thinkers' intent on maintaining just wars.

Coker and other security thinkers imagine the future of war to be fast paced and technologically advanced, and that this pace and advancement will escalate as time progresses. This will move war beyond the cognitive and physiological capabilities of human warfighters, until humans are removed from the battlefield all together (Coker, 2015: 53–68). Scharre offers a number of examples of the tensions and problems that are emerging from wars that are beginning to take place in this way, many of which revolve around accidents. Accidents are not unique to modern and emerging technologies in war. They are always present in war, and everyday life, both because of technological malfunction, and human error. However, Scharre argues that the accidents present in the contemporary and future battlefield, because of technology, stand out from those before specifically because of the speed and complexity of the technologies at play. For instance, Scharre discusses the phenomenon of the 'Runaway Gun', an automatic weapon that has malfunctioned and so does not stop firing when the person operating it releases the trigger. This leads to the gun continuing to fire uncontrollably; resulting in unplanned damage, death, and friendly fire (Scharre: 190–1). As such, Scharre explains that:

> If there was no human to intervene, a single accident could become many, with the system continuing to engage inappropriate targets until it exhausted its ammunition. 'the machine doesn't know it's making a mistake,' Hawley observed. The consequences to civilian or friendly forces could be disastrous.
>
> (Scharre: 191)

This is especially true when such technologies are complex, fast paced, and widely networked. Here, such malfunctions would not be isolated to one unit of AI weaponry (such as a single gun), but rather could affect the rest of the network, cascading though a single system, or affecting other connected systems (i.e. other guns, monitoring systems and command and control software). John Borrie calls this situation an 'incident of mass lethality' (Borrie in Scharre; 193). This is a worrying prospect, especially when considered through the lens of military chaplaincy. The potential lethality and destruction of AI accidents causes a host of problems, pastorally and practically, which the military chaplain will need to respond to, on top of those created by the introduction of AI in of itself (Wilkinson-Roberts, Forthcoming). However, I am concerned with the fact that at the point of a malfunction, *jus in bello* can no longer be maintained. For the AI agent is unable to prevent damage to allies, enemies, and civilians. The gun can no longer achieve proportionality or discrimination, and so cannot act justly or unjustly. In this context, civilian casualties are no longer a matter of cognitive processing, but of whether the gun happens to fire at them or not while on its rampage. Here, *jus in bello* is up to chance, maintained only by luck and coincidence.

At this point, maintaining morality is no longer the responsibility of the chaplain, for there is no choice being made here. Instead, the JWT ceases to operate in these conditions. Of course, this does not nullify the role of chaplaincy. Chaplains will inevitably have to deal with the moral choices and pastoral needs of the humans surrounding the runaway gun. Most relevant here are the responses of human warfighters having witnessed an otherwise unjust act take place without being able to prevent it. Unlike conventional unjust acts, where a person chooses to act badly and is prosecuted accordingly. An AI accident is not someone's choice, but a mistake. Thus, it cannot be thought of like the unjust actions I explained in the first section. Witnessing such devastation is likely to create abundant pastoral and moral challenges for the personnel surrounding the incident, as well as for military decision makers more broadly.

I will return to this point in a moment, but first, it is necessary to discuss a slightly different kind of accident. That is, an accident without a malfunction. This kind of accident is unique to technology, as it is an accident produced by technology functioning as it is intended, but in a way which is unhelpful for – and unpredicted by – human tacticians (and technicians). Scharre uses the example of a 'Flash Crash' to explain this, using an instance from online auctioning to warn readers about AI non-malfunction accidents (Scharre: 199–210). In this example, two AI bookselling programmes took part in a 'runaway price war', each increasing their costs to compete with the other. However, the AI did not stop doing this once the costs got out of hand with the cost of the book in question reaching $23,698,655.93 (plus shipping) before a human stopped the AI (Scharre: 205). The problem was not just that the AI did not know to stop bidding when it reached a certain price, but that the process of bidding, meeting a bid, a re-costing happened at such a fast pace compared to when humans bid. Scharre reflects that:

> Stock trading [and book selling] is a window into what a future of adversarial autonomous systems competing at superhuman speeds might look like in war. Both involve high-speed adversarial interactions in complex, uncontrolled environments. Could something analogous to a flash crash occur in war – a flash war?
>
> (Scharre: 207)

He goes on to speculate about how AI might operate in conditions where decision-making is subtle and loaded, such as stand-offs where escalation is not guaranteed. Like Payne, Scharre is concerned about whether AI would effectively operate as humans expect in these conditions (Scharre: 208). In the Flash Crash example, the AIs did not malfunction. They functioned correctly according to their programming. However, they did so to such an extent, and at such speeds that it resulted in a financial accident. Unlike the runaway gun example, the accident here is not a removal of agency and control resulting in damage, but rather a correct running of agency (as determined by human programmers) that results in damage.

It is unclear what this might look like in a military context. However, as Zweibelson and Payne highlight in their works, the cognitive dynamics of AI is significantly different to that of humans. This is likely to lead to mistakes in programming and application like those in Scharre's Flash Crash example (Zweibelson: 29–34; Payne: 219–221). Resulting in situations where technologies behave correctly but in ways that humans do not anticipate, resulting in damage to bodies, buildings and military assets. Similar to the runaway gun example, it is difficult to deal with this within the just war framework, as the AI is not making a moral miscalculation – or even malfunction – to avoid or prosecute. This may lead to a situation where a technology designed to make proportional, discriminatory and limited actions may do so to such an extent that it results in unjust activity. This is especially likely due to the complexity of just war calculations which typically require human judgement to make effectively (Payne: 219–235).

This is typically the point where we cut our losses (theoretically) and suggest that the military requires intentional plans to mitigate mistakes and malfunctions, and to utilise complexity theory to ensure we imagine the ways technology might understand and respond in battle situations. Thus, mitigating the challenges that accidents might create on the battlefield. However, to do so would misunderstand both the scope of technological accidents and their challenges for the JWT.

According to Paul Virilio, accidents are not aberrations, but are instead inevitably bound to technological innovation (Virilio, 1998, Virilio, 2006). With every new technology, comes new accidents from that technology:

> When you invent the ship, you also invest the shipwreck; when you invent the plane, you also invest the plane crash; when you invent electricity, you invent electrocution… Every technology carries its own negativity which is invented at the same time as technological progress.
>
> (Virilio, 1999: 89)

This also means that avoiding the accident, though theoretically possible, will lead to other kinds of accidents as new technological components, and systems are added, and with them new accidents. We therefore, cannot cut our losses, for they cannot be avoided in the first place. We may work to avoid a runaway gun, but in doing so stumble into a different, and inevitable, accident.

In a future operational environment, this is compounded by the interconnectedness of technological systems, their often-lethal capabilities, and their fast pace. This means that military AI carries with it a host of inevitable accidents, each with catastrophically damaging potential. For Virilio, this is a truth of technological innovation and globalisation which disrupts our now typical optimism regarding the 'technological fix' to technological, social and political problems (Lacy, 2014: 105–6). As technology becomes more complex and commonplace, so do accidents, and we cannot – despite our best intentions – innovate away these accidents and their consequences. The technological fix, therefore, is a fallacy of modernity (Lacy, 2014: 105–9). Virilio is often critiqued for being too negative and apocalyptic about the future. However, it appears that the caution that Virilio asks of us – and his suspicion of technology and their surrounding discourses – mirrors that expressed by the other authors covered in this chapter. For them, AI promises much, but also brings challenges. Some of which may contradict the benefits of AI altogether. This may indeed seem like an apocalyptic prediction, however, through the lens of chaplaincy, it is a necessary one. Chaplaincy is not just about dealing with the present needs of military personnel, but also about thinking of the future, scouting out what challenges may be coming. This is why Chaplains are involved in moral leadership. They do not just react to the immoral choices a warfighter might make, but prepare the fighter to make good choices in the first place.

Virilio may be doing the same thing. Imagining the future so that we can adequately prepare ourselves for the challenges to come. Through a chaplaincy lens, there is a question here about how the just war tradition will fair in an environment when unjust events are inevitable due to the accidents of technology. Is it possible to have a just war whilst AI threatens such catastrophic mistakes? How can we justify going to war if such a war will mean a lack of proportionality, discrimination and limitation when these technologies malfunction? For, accidents are more likely, and more damaging, the more sophisticated and ubiquitous emerging technology becomes. Returning to the front lines, how might soldiers deal with witnessing the morally unjust (but accidental) actions of their AI counterparts? What is needed of chaplains in these circumstances?

These questions are difficult to answer ahead of time, and go beyond the scope of this short chapter. We may want to look to history for some of the answers. Accidents in war are not new, as the runaway gun and flash crash examples show us. Nonetheless, what viewing the future of war though a chaplaincy lens shows us, is that military operations are likely to be fraught with challenges, which we must reckon with, lest we be surprised by some catastrophic existential challenge like those articulated by this chapter.

CONCLUSION

In this chapter, I have argued that the use of AI creates circumstances that make the just war theory vulnerable. In some ways, AI invites unjust action – whether by human personnel or the AI itself. Additionally, the likelihood and scale of accidents from AI and its related technologies creates an environment where unjust action (in the form of malfunctions and mistakes) are always possible. This leaves the Just War Theory wounded. For, military personnel, decision-makers and chaplains are likely to find themselves in a context where just action in war will be more difficult and complex than ever before.

It is very likely that AI-powered technologies will continue to be researched, developed and deployed throughout the military, due to their cognitive power and speed. For me, it is not the specific application of AI technology that is important, but the characteristics and capabilities of the technologies themselves: as artificial, and intelligent. This is coupled with the ways that the battlefield and warfighting transform around them, compounding their challenge for theologians and security studies thinkers. This chapter has explored a number of these challenges though the specific lens of the military chaplain. Chaplains are vital moral leaders, providing education and oversight to ensure that personnel are equipt to make *jus in bello* (just activity in war) compliant choices in the theatre of war. By exploring the challenges of AI through the lens of the chaplain, this chapter exposes the subtle ways that AI affects warfare and warfighting in a pastoral and metaphysical sense, highlighting otherwise overlooked aspects of AI's consequences. In particular, I used the work of key contemporary scholars to articulate some of the ways that AI affects actions in war, whether they are intentional actions or accidents. These include how: (1) AI can create space for moral recklessness and contradiction; (2) accidents can nullify *jus in bello* all together; (3) accidents can puncture our ability to predict and understand AI; and, (4) AI itself changes the character of warfare, reducing its anthropocentricm and thus endangering the JWT. Each of these create unique challenges for Military Chaplains, but also require a greater emphasis on the chaplain as a moral leader, who is aware of the kinds of challenges that AI can cause, so to effectively support their personnel. Likewise, chaplains will be required to stress the importance of moral courage, vigilance and bravery, so to protect against the moral recklessness AI might invite. Lastly, Chaplains must remain vigilant of impending accidents (from malfunction and misunderstanding) so to best prepare for the moral and pastoral challenges of the future.

This may seem like a bleak prediction. However, despite the challenges I have outlined in this chapter, I am not arguing that AI is a hopeless invention, or that its dangers absolve it of its benefits. Rather, I argue that we cannot – in good conscience – invest in the benefits of AI without taking seriously its inevitable consequences and accidents too. We must continue to invest in a more just and effective future, but we cannot ignore the challenges that make up our invention's shadows; not if we want our activities in war to be just.

REFERENCES

Ahmad, A. (2006). The Evolution of Just War Theory in Islamic Law: Texts, History, and the Purpose of "Reading". *American Foreign Policy Interests*, *28*(2), 107–115. https://doi.org/10.1080/10803920600663501

Beauchamp, T. (2019). *Going Stealth: Transgender Politics and U.S. Survaillance Practices.* Duke University Press.

Buolamwini, J. (2023). *Unmasking AI: My Mission to Protect what is Human in a World of Machines*. Random House.

Chamayou. G. (2015). *Drone Theory*. Penguin Books.

Clausewitz, C. (2012). *On War, Volume 1*. Duke Classics.

Coker, C. (2015). *Future War*. Polity Press.

Fairclough, G. (2019). The Ethical Challenges of Information Warfare: Nothing New. In T. S. Mallard & N. H. White (Eds). *A Persistent Fire: The Strategic Ethical Impact of World War I on the Global Profession of Arms*. (pp. 165–185). National Defense University Press.

Feuer, J. C. (2022). The Korean Buddhist Military Chaplaincy and Modern "State-Protection" Buddhism: A Study of the Mass Military Faith Promotion Movement. *Journal of Korean Religion*, *13*(2). https://go.gale.com/ps/retrieve.do?tabID=T002&resultListType=RESULT_LIST&searchResultsType=SingleTab&retrievalId=c6d43624-9382-4ac2-b15f-be126c160e25&hitCount=1&searchType=AdvancedSearchForm&currentPosition=1&docId=GALE%7CA724286386&docType=Article&sort=RELEVANCE&contentSegment=ZONE-MOD1&prodId=AONE&pageNum=1&contentSet=GALE%7CA724286386&searchId=R2&userGroupName=unilanc&inPS=true

Fisher, P. (2013). The Robotisation of War: An End of Military Values? In A. Todd (Ed). *Military Chaplaincy in Contention: Chaplains, Churches and the Morality of Conflict* (pp. 83–92). Ashgate.

Fyfe, S. (2017). The Challenge to the Law of War by Islamic Jihad. In L. May. (Ed). (pp. Chapter 16). *The Cambridge Handbook of the Just War*. Cambridge University Press. https://doi.org/10.1017/9781316591307

Lacy, M. (2014). *Security, Technology and Global Politics: Thinking with Virilio*. Routledge.

Lee, P. (2019). *Reaper Force: The Inside Story of Britain's Drone Wars*. John Blake Publishing.

Lewis, J. (2022). *Jewish Chaplaincy in the British Armed Forces: Captains of the Souls of Men 1892–2021*. Vallentine Mitchell & Co.

Michalowski, I. (2015). What is at stake when Muslims join the Ranks? An International Comparison of Military Chaplaincy. *Religion, State and Society*, *43* (1), 41–58. https://doi.org/10.1080/09637494.2015.1021189

MOD, (2021). Defence in a Competitive Age. (CP 411). *Crown Copyright*. https://assets.publishing.service.gov.uk/media/6063061e8fa8f55b6ad297d0/CP411_-Defence_Command_Plan.pdf

Moyn, S. (2022). *Humane: How the United States Abandoned Peace and Reinvented War*. Verso.

Payne, K. (2021). *I, Warbot: The Dawn of Artificially Intelligent Conflict*. Hurst & Company.

Scharre, P. (2018). *Army of None: Autonomous Weapons and the Future of War*. Norton Paperback.

Snape, M. (2008). *Clergy Under Fire: The Royal Army Chaplain's Department*. The Boydell Press.
Swift, C., Cobb, M. & Todd, A. (Eds). (2015). *A Handbook of Chaplaincy Studies: Understanding Spiritual Care in Public Places*. Ashgate Publishing (Routledge).
Virilio, P. (1998). *The Virilio Reader*. J. Der Derain (Ed). Blackwell.
Virilio, P. (1999). M. Cavaliere (Trans.). *Politics of the Very Worst: An Interview with Philippe Petit*. Semiotext(e).
Virilio, P. (2006). M. Polizzotti (Trans). *Speed and Politics*. Semiotext(e).
Wilkinson-Roberts, K. (Forthcoming). *Pastoral Security in the Future Battlefield*. PhD Thesis. Lancaster University.
Winfield, T. P. (2022). Chaplaincy Work and Preparation across Sectors. In W. Cadge & S. Rambo (Eds). *Chaplaincy and Spiritual Care in the Twenty-First Century: An Introduction*. (pp. 32–57) The University of North Carolina Press.
Zweibelson, B. (2023). Part 1: The Singleton Paradox – On the Future of Human-Machine Teaming and Potential Disruption of War Itself. *Journal of Advanced Military Studies 14*(1), 11–46. https://doi.org/10.21140/mcuj.20231401001

9 Wargaming and the Just War

David J. H. Burden

'Wargames save lives' declared leading US professional wargame designer Matt Caffrey (Caffrey, 2018). This chapter examines the potential truth of that statement within the context of Just War Theory, and particularly with regards to the increasing impact of Artificial Intelligence (AI) on the battlefield and in wargaming itself. The chapter argues that wargaming can make a significant contribution to the examination of just war considerations, both within the professional military community and beyond. The chapter begins by describing what constitutes wargaming and how it is used by the military and others. The chapter then examines how AI could be used in wargaming, and how urban warfare, and by extension urban wargaming, provide the *in extremis* case for studying just war and wargaming. The main body of the chapter then examines the role of wargaming, and AI in support of war and wargaming, in each of three elements of a just war analysis – *jus ad bellum, jus in bello* and *jus post bellum*. The chapter concludes by arguing the case for a broader use both of wargaming in considering just war issues, and in including just war issues within wargames, and by highlighting the need for just wargames to support just wars.

WARGAMING

There are many definitions of wargaming, but for the purposes of this chapter the definition in the U.S. Department of Defence (DoD) *Joint Publication 5-0, Joint Planning* will serve:

> Wargames are representations of conflict or competition in a synthetic environment, in which people make decisions and respond to the consequences of those decisions.
> (US Department of Defense, 2020)

A Very Brief History of Wargaming

In its current form, wargaming developed under the Prussian military in the nineteenth century (Caffrey, 2019), and was used extensively by the US Navy between the wars (Appleget et al., 2020), and by all sides during World War 2 (Perla, 2011). Although manual wargaming continued after World War 2 it was for several decades eclipsed by computer-based wargames (Perla, 2011), and today both manual and computer forms of wargaming are in regular use (Mouat, 2022).

 DOI: 10.1201/9781003440253-9

FIGURE 9.1 A typical professional manual wargame – the British Army's RCAT.

Source: Author's photo.

Wargaming Structures

The typical manual wargame consists of a map, a set of counters or blocks, and a set of rules. Figure 9.1 shows a typical professional wargame.

One way of categorising wargames is by how free (i.e. they have minimal rules and an umpire to adjudicate actions) or rigid (i.e. they have a detailed rulebook) they are. Another consideration is whether they are open (where players can see all the pieces and know all the capabilities – such as in chess) or closed (where significant amounts of information might be hidden from the players – as in poker). Wargames can sit anywhere in a space defined by these measures (Figure 9.2). The matrix game, which is a fairly free wargame, usually with some hidden information, and supports structured discussion and argument is of particular use where there are significant unknown or human factors and so are of particular interest to this current study (Curry et al., 2011).

How Wargaming Is Used

The professional community differentiates between wargames for training, *'to provide decision-making experience'* and wargames for analysis, *'to provide decision-making information'* (Longley-Brown, 2019, p. 48). A further typology could extend to include:

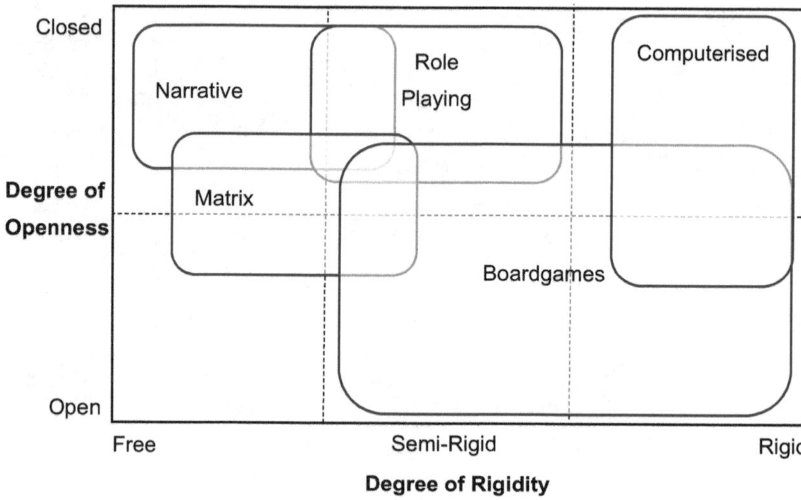

FIGURE 9.2 The free to rigid wargame spectrum against open or closed games.

- wargames as entertainment;
- wargames as edutainment (a balance of learning and fun) (Silva Simões & Ferreira, 2011); and
- wargames as education (where learning is the main goal, and enjoyment more of a motivational factor).

One important point to note is that wargames are not considered to be predictive tools, in the sense that they will give an absolute answer to a question. Rather they can help to outline a landscape of possibilities, and, as described by Mouat (2022), provide some insight into what outcomes may be possible, or even more likely. In many wargames, even professional ones, rigor and plausibility are more useful reference points, the purpose often being to stimulate player decision-making for insight, rather than making predictions on outcomes (Webber, 2021).

AI in Wargaming

Figure 9.3 illustrates the typical roles that need to be fulfilled in a professional wargame. Any of these could *potentially* be fulfilled by either an AI or a human – or a human-machine team, or a human supported by an AI tool. AI implementations are at early and varied levels of maturity. For it to remain a wargame, rather than a simulation, at least one of the players needs to be a human.

A 2023 study (Knack & Powell, 2023) from the Centre for Emerging Technology and Security (CETaS) at the Alan Turing Institute identified 4 main activities within a wargame, and within those the different areas where AI could make an impact, as shown in Figure 9.4.

CETaS identified the potential role of AI in both narrow and specialised applications, such as the generation of synthetic social media messages, and high-risk high-reward AI applications in game execution and analysis. Challenges identified included:

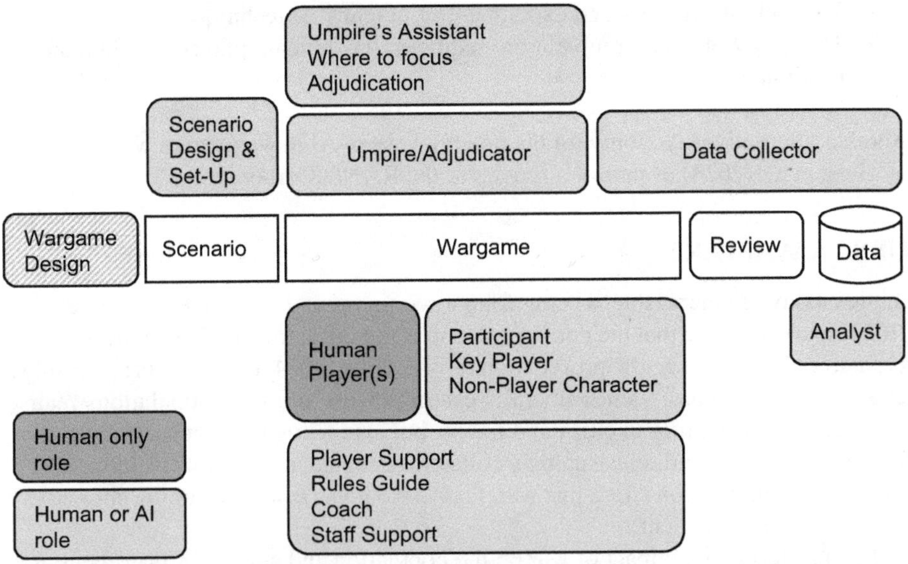

FIGURE 9.3 Human and AI roles in a wargame.

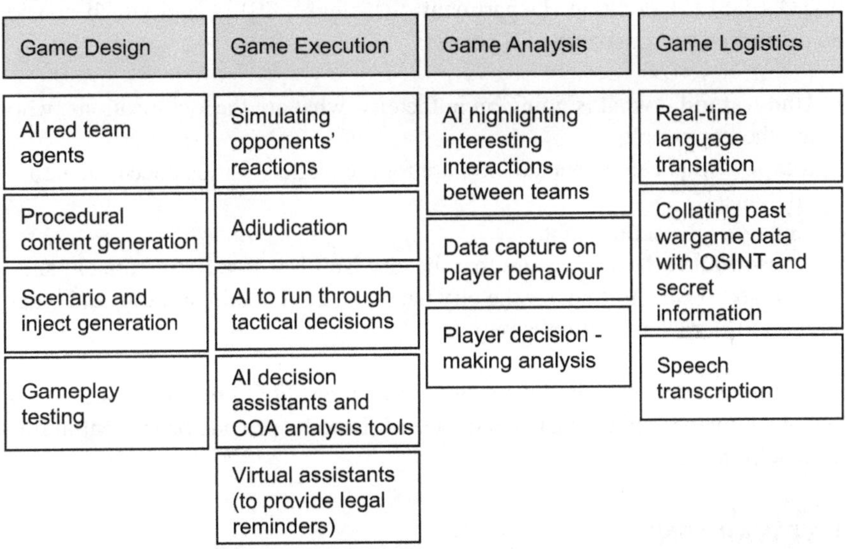

FIGURE 9.4 Long-list of AI use cases for wargaming – after Knack & Powell, 2023.

- Developing a valid methodology and model, especially when trying to model irrational human actors and/or to capture the entire environment which is affecting decisions;
- Over-trust in the outputs of the AI;
- The ethical implications of making decisions as a result of the wargames which have been in part the output of an AI;

- The lack of causality and explainability in many AI techniques;
- The psychological preparation required for human players in human-machine teams.

Another alternative taxonomy for the potential use of AI in wargaming was produced by Geist et al. (2024) as part of a report for the RAND Corporation.

URBAN WARFARE

Battles today are increasingly being fought within urban environments, where King (2021) and others see that the density of people and infrastructure, and concepts such as smart cities challenge the prosecution of war. Hills (2004) describes urban conflict as a 'liberal dilemma' – since liberal western powers and their populations realise the necessity of fighting in cities and towns, but also recognise that such fights may lead to significant civilian casualties, collateral damage and loss of livelihood. This has obvious implications for a just war. Figure 9.5 gives some idea of the complexity of the urban environment.

In considering the impact of war on the population and a city's infrastructure it is vital to widen the scope of any military planning to consider what can be learnt and done before the fight, and what may need to be done afterwards. This is encapsulated in NATO's USECT doctrine (Department of Defense, 2013), and which considers urban operations as consisting of 5 phases:

- **Understand** – what is going on in the city, what are the key locations, who are the key people;
- **Shape** – what can be done to prepare for the battle – for instance evacuating civilians;
- **Engage** – the actual fight;
- **Consolidate** – how to ensure that the city isn't lost in a counterattack; and
- **Transfer** – how to transfer the city back to a civil administration, and what support in reconstruction is required.

Urban conflicts offer new challenges as to how they should be wargamed (Burden, 2023a) and how to keep war just, and the USECT model provide one potential framework to help in this.

A JUST WAR LENS

A traditional view of Just War Theory (Walzer, 2015) sees it as providing the moral basis for international law as applied to armed conflict. For the purposes of this Chapter, the three-phase model based on Bass (2004) and Stahn et al. (2014) will be used:

- *jus ad bellum* – is it just to go to war;
- *jus in bello* – is the war being prosecuted in a just way; and
- *jus post bellum* – is the peace implemented and managed in a just way.

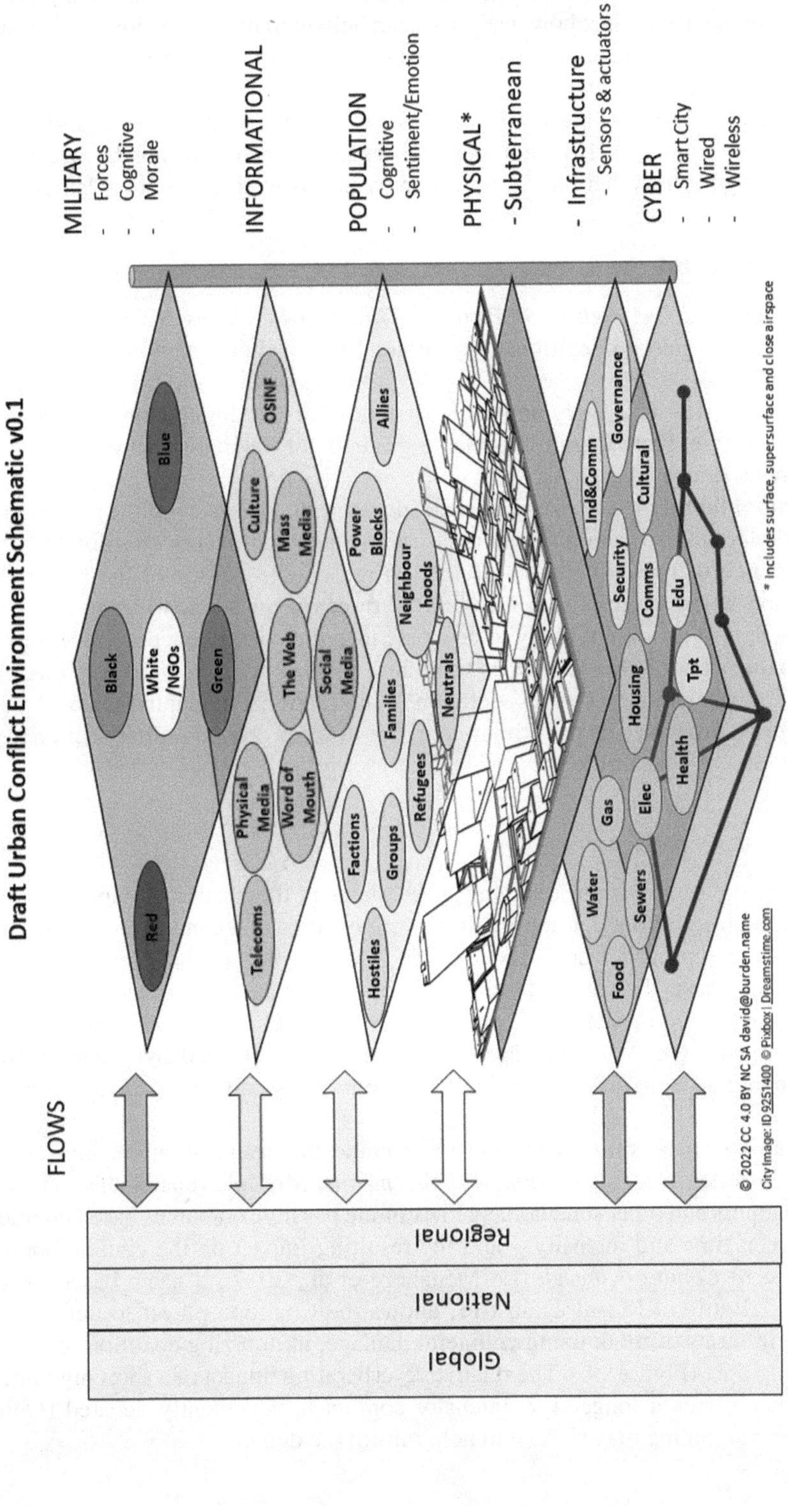

FIGURE 9.5 The urban conflict environment.

Using this three-phase model as a structure, the Chapter will consider each phase in turn, and for each examine: how wargames can help with making actions (more) just, how AI is impacting the conduct of the actions, and how AI can help improve the wargaming of those actions.

Whilst *jus ex bello*, bringing wars to a just end, as proposed by Mollendorf (2008, 2015) and Colonomos (2015) is not separately examined here, many of the discussions of wargaming for the other phases would apply just as well to this consideration.

JUS AD BELLUM

The UN *Report of the High Level Panel on Threats, Challenges and Change* identified 'five basic criteria of legitimacy' for going to war (United Nations, 2004):

- **Last resort** – Has every non-military option for meeting the threat in question been explored, with reasonable grounds for believing that other measures will not succeed?
- **Proportional means** – Are the scale, duration and intensity of the proposed military action the minimum necessary to meet the threat in question?
- **Balance of consequences** – Is there a reasonable chance of the military action being successful in meeting the threat in question, with the consequences of action not likely to be worse than the consequences of inaction?
- **Seriousness of threat** – Does the threat justify the use of military force, including imminent serious violation of international humanitarian law?
- **Proper purpose** – Is the primary purpose of the proposed military action to halt or avert the threat?

WARGAMING *AD BELLUM*

In terms of the *last resort* wargaming gives planners the chance to explore a wide variety of options before committing to the use of force and can help to either identify non-force approaches or to generate the evidence base to show that other measures will not succeed. Free wargaming techniques such as matrix games, and related activities such as confrontation analysis described by Howard (1999), can be useful in exploring the grey-zone/sub-threshold alternatives described by Brands (2016) and Carment and Belo (2018) – including diplomatic, economic, cyber and terrorist operations.

Wargaming an operation can help to examine the scale, duration, impact and intensity in order to assess the *proportional means*. Multiple runs of the game can enable planners and other stakeholders to examine possible outcomes based on trade-offs between time and intensity – and the resulting impact on the civilian populations. *We Are Coming Nineveh* (Le Ménahèze et al., 2017), a game based on the 2016–2017 Battle of Mosul against IS, allows the Coalition player to define their objective in terms of minimizing collateral damage, minimizing coalition losses, or minimizing time (Figure 9.6). The relative reverberating impact of a short high intensity conflict verses a longer low intensity conflict is vehemently debated (Ohlin, 2017) and wargaming may be able to help inform the debate.

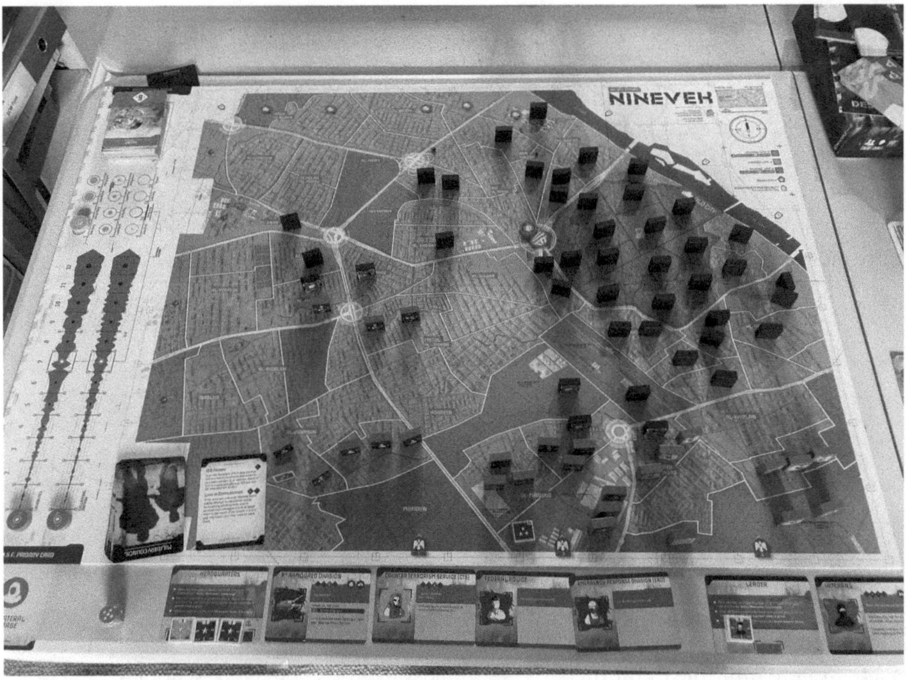

FIGURE 9.6 We are Nineveh.

Source: Author's photo.

In terms of the *balance of consequences* the wargame can again provide evidence to help gauge the likelihood and cost of success. As mentioned, wargames should not be seen is predictive, but running multiple iterations should generate a probability map of the outcome space. The wargame can also provide information on the possible reverberating effects of the action, and a matrix game could explore the consequences of inaction. The *seriousness of threat* is a far more subjective judgement, but even here a matrix game could help to explore the possible evolution of a situation. Whether *proper purpose* will be met is very much dependent both on detailed sight of the military plans, and on the plans being a true representation of what is intended, or likely, to happen on the ground. Wargaming a set of plans or a scenario can provide a deeper understanding and insight into them than just reading, and may also identify areas where there is more risk of *proper purpose* being exceeded – particularly when gamed with a matrix approach where the game mechanics place less control on the players and give them greater scope of action – and where player briefings might highlight existing tensions and misalignments that could lead to such problems.

Whilst military and professional wargames can be hard to identify and obtain information on, even some hobby games deal with *jus ad bellum* issues, and so could be useful to help educate people on some of the basic issues. For instance, in both *Twilight Struggle* (Gupta & Matthews, 2005) about the Cold War and *Flashpoint: South China Sea* (Buchanan, 2022) on the current issues in that part of the world, players are trying to grow influence and prestige but without escalating to open war.

AI and *Ad Bellum*

The *ad bellum* phase of any conflict is increasingly becoming one of grey-zone operations, where, as described by Azad et al. (2023), the conflict is always ongoing, the only issue is whether it has broken out into open warfare. This might fundamentally challenge the notion of *jus ad bellum*. In the confrontations are always *in bello*.

Some key areas in which AI is, or is expected to be, used in grey-zone confrontations include:

- The running of bot farms and generating fake social media accounts and content for influence and disinformation operations;
- The attack and automated computer-speed defence aspects of cyber-warfare;
- The modelling of population and personality intent and behaviours for best targeting of grey-zone operations;
- The detection of grey-zone activity by a hostile actor, and decisions on the best responses.

AI, Wargaming and *Ad Bellum*

In each of the three phases under discussion it is tempting to think that an AI could just complete multiple iterations of a wargame and advise on the best course of action. However, as discussed above, a wargame without a human player just becomes a simulation, and whilst that is still a valid operational analysis approach, it is suggested here that this should be done alongside human wargames in order to better factor in the human elements of war.

Free wargames are useful in considering the *ad bellum* phase, but one of the challenges is in getting players to reflect the mindsets of potential opponents. Examples of where this has been tried with human players include the Japanese in 1942 who had recently repatriated naval defence attaches and other staff to stay in a US mindset when playing the US forces in Japanese wargames (Perla, 2011), and the current Chinese Navy is actively establishing its own immersed Blue Teams (Martinson, 2024). As AI develops it could become possible to create digital-twin AI personalities modelled on not just generic people and adversaries as described by Barachini and Stary (2022) but, as suggested by Vakas et al. (2001) and Tandon & Mehra (2023), on very specific commanders and politicians. Hunter & Bowen (2023) consider that current approaches to AI may not be well suited to this, however there is no certainty that this will remain the case. As a result, the line between simulation and wargame is likely to become increasingly blurred as the capability of human-similar AIs develop.

An alternative AI approach is that of agent-based modelling (ABM) as described by Kerr et al. (2021). The *ad bellum* phase has multiple stakeholders from international organisations, non-governmental organisations (NGOs) and corporates, to politicians and individual citizens. In ABM all of these are represented at an individual level, each given their own behaviours, and then left to respond to not only the inputs from the wargame but also to the behaviours of the other agents. Humans still play the key roles in the wargame, but the behaviours of these other elements modelled through ABM hopefully generate a more realistic output. For a population this may be at a gross level in terms of changing sentiments, confidence or fears, or at a more individual level in terms of moving to safer parts of a country, withdrawing funds from banks, generating overloads on the communications networks, or posting messages to social media – all of which may then influence other people/agents to do the same thing, and which may then have knock-on effects on infrastructure management, business and country economies, and of course the politicians. Such modelling, within or outwith the context of a wargame, could help in better understanding the implications of any decision to go to war.

Whilst many grey-zone actions may be implemented by AI there is no reason why they have to be modelled by an AI in a wargame. For instance, the RAND Grey-Zone wargame (Pettyjohn & Wasser, 2019; Wasser et al., 2019) was a manual wargame which used cards to represent different grey-zone actions, many of which would be AI enabled in real-life, and this approach is taken by many other modern games.

JUS IN BELLO

For *jus in bello* the typical criteria considered about whether the actual execution of the war is just include:

- **Discrimination/Distinction** between combatant and non-combatant, including in the causation of collateral damage as described by Bica (1998);
- **Proportionality** of the response as considered by Van Steenberghe (2012);
- No means *mala in se*, i.e. no evil weapons as discussed by Dige (2017); and
- **Military Necessity** as examined in Patterson & LiVecche (2023).

Wargaming in bello

Most wargames are probably conducted about the *in bello* phase. Such wargames will typically start as the two sides are about to engage in combat, and end once there is a clear winner, or objectives have been achieved. These wargames only cover the Engage phase of the USECT model, but better Understanding and Shaping before the Engage phase can help in reducing civilian harm and mitigating reverberating effects, and the Consolidation and Transition elements can similarly aid in the tactical or strategic *post bellum* phase. Rigid wargames are probably more relevant here than in the other phases, but free wargames continue to have a role, and a combination of both perspectives is probably desirable. For *discrimination* or *distinction* to be modelled requires that the wargame represents the civilian population (to give the players something to distinguish between), possibly have some level of hidden information (so it's not immediately apparent whether a building is occupied by civilians or the enemy), and that collateral damage is also modelled or tracked. Unfortunately, most wargames do not represent the civilian population to any meaningful degree (Burden, in press).

About the only hobby boardgame/wargame which currently make civilians the focus of the game is *This War of Mine* (Oracz & Wiśniewski, 2017). A matrix game with a similar focus could give similar benefits – enabling planners to better understand the plight of those who will be caught up in the fighting. A well-briefed team playing the civilian population would also help give the players of a large wargame or Command Post Exercise (CPX) a better understanding the impact of their actions on the civilian population. Modern wargames such as *We Are Coming Nineveh* (see Figure 9.6) and the author's own *City & CEMA* (Burden, 2023b) use blocks to provide hidden information. If a side attacks these blocks without first identifying if they are enemy troops or civilians then the game can generate a suitable penalty, such as points on an adverse opinion track. The impact of civilian deaths can even be amplified by the presence of press or media, identified as counters or as even cards.

Collateral damage is slightly more commonly tracked in games than civilian casualties, often in the form of the generation of rubble, and again this might contribute to an adverse opinion count. However, no urban game yet examined really considers the reverberating effects of collateral damage (e.g. hitting a power station stops sewerage pumps which leads to more health problems which puts greater strain on a hospital which has already lost power and pumped water) or the longer term impacts on livelihood, although *City & CEMA* (Figure 9.7) has some sense of reverberating effects, as well as modelling all of the USECT phases.

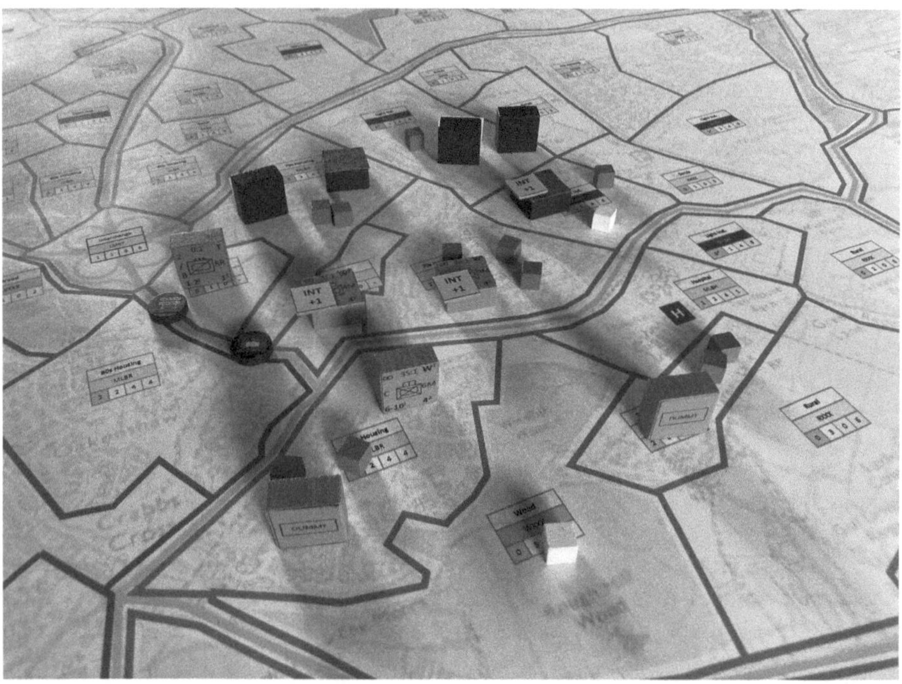

FIGURE 9.7 City & CEMA, with grey cubes showing collateral damage and white cubes the presence of civilians.

Source: Author's photo.

To assess *proportionality* a wargame really needs to track both the achievement of each side's military objectives, and the reverberating damage that they cause in achieving those objectives. A comparison between the two (or an absolute ceiling for the latter) could help assess the proportionality of an operation, and different tactics tried in multiple play-throughs in order to identify a best (or more probably a least worse) strategy. Such a proportionality metric (or even just the tracking of civilian/collateral damage or adverse opinion) could also be used to trigger game actions such as a UN ceasefire, or an ally being drawn in on the side of the beleaguered civilians. For instance, Brian Train's *We Are Coming Gaza* variant for the *We Are Coming Nineveh* game (which was written between the Hamas attacks and the Israeli Defence Force (IDF) ground offensive in October 2023 in the spirit of game-based journalism) features an optional rule where the rolls for a UN imposed ceasefire start earlier the more collateral damage the Israelis cause.

In terms of *mala in se*, the use of flame and gas is of particular interest. Educational wargames can show how such weapons have been used in the past – and what their effect was. Analytical games should provide a safe opportunity to explore the pros and cons of these weapons on the modern battlefield. Flamethrowers were commonly used in World War 2, and so many games depicting that period include these weapons. There is a danger though that they just become a numeric bonus in combat, and so their human implications are not fully considered. One of the perceived values of flame weapons is that they can force enemy troops out of buildings whilst keeping the core structure of the buildings intact – so reducing collateral damage (US Army, 1964, p. 31–50). A game might reduce collateral damage if flame is used, but adverse opinion would certainly increase if there were civilians in the building, and even its use on the enemy might cause some increase in adverse opinion. Three of six wargames examined on the 1968 Battle of Hue included the use of CS Gas (since it was heavily used by the US troops in the battle), with varying negative consequences – so the player has a trade-off in terms of whether or not to use it (Burden, 2023a).

One issue to beware of is that, in writing the rules for the effects of the weapons the game designer will almost certainly be bringing in a set of biases, and these need to be surfaced and addressed if the game is not to be a self-fulling prophecy, one of the worst forms of wargame pathology as Weuve et al. (2004) indicate.

Evaluating *military necessity* probably requires a degree of expert judgment from an umpire or observer, identifying which actions a player takes are necessary to achieve their objectives. Given the tight focus of many wargames the designer may need to ensure that the wargame actually offers players the chance to take unnecessary actions (although of course not described as such).

AI AND *IN BELLO*

The role of AI within the *in bello* phase of warfare is a under significant discussion at the moment – as this book attests. AI can be expected to contribute to:

- Surveillance and targeting systems;
- Intelligence analysis;

- Autonomous platforms such as Uncrewed Ground Vehicles (UGVs) and Uncrewed Aerial Vehicles (UAVs);
- Course of Action planning;
- Electronic and cyber warfare; and
- Information warfare.

AI, Wargaming and in bello

Existing wargaming approaches can be used to model many of the effects of AI, given the level of abstraction in most manual wargames – although these are often areas which manual wargames do not yet cover. One area where AI can assist in *in bello* wargaming is in the representation of the civilian population. As discussed, the civilian presence in a wargame is often sketchy at best or missing altogether. However, AI can be used to help create the ambience of conducting operations within and amongst the people, so that commanders can get immediate and natural feedback from the civilian population as to the impact of their operations. One example of this is in how an AI system was developed by Daden Limited[1] for MOD to provide simulated social media feeds from the inhabitants of a city which was being fought over between two sides in a wargame (Burden, 2022). Another junction of wargaming and AI is that when seeking to evaluate the ethics of using lethal autonomous weapons in urban warfare planners need to have a sense of their operational impact and the collateral damage that they might cause – and wargaming is one way of providing this missing information as suggested by Allen and Baker (2022).

As military decision-makers increasingly base their decisions on the recommendations of AI-driven software then perhaps wargames need to include this AI element and explore the varied issues of human-machine and human-autonomy teaming. The line between simulation and wargaming may also become further blurred should having the human in the loop in the actual warfighting becomes less common, and the transition is made to more of a human-on-the-loop model with just a power of veto.

JUS POST BELLUM

The idea of *jus post bellum* as a set of principles to monitor the '*moral propriety of concluding a war*' was described by Shuck (Schuck, 1994), although the concept dates back to at least Kant's *Right after War (Recht nach dem Krieg)* (Stahn, 2007). The case to consider *jus post bellum* alongside *jus ad bellum* and *jus in bello* has been widely supported, such as by Bass (2004) who sees it as necessary to complete the theory of just war and Iasiello (2004) who considers it a necessity in the light of the post-invasion experiences in Afghanistan and Iraq. Bellamy (2008), however, considers it premature as *jus post bellum* has scope beyond that of the Just War tradition, and involves multiple actors. Bass (2004) identifies three key questions to consider:

- What obligations are there to restore the sovereignty of a conquered country and what limitations do these obligations impose on states' efforts to remake the governments of vanquished countries?

- What are the rights and obligations that belligerent states retain in the political reconstruction of a defeated power? Are these rights limited to the reconstruction of genocidal regimes, or can a case be made for the political remaking of less dangerous dictatorships?
- What obligations might victorious states have to restore the economy and infrastructure of a defeated state? And conversely, do victorious states have a right to demand some kind of reparation payments from defeated states who were aggressors in the concluded war? (p.385)

Iasiello (2004) sees *jus post bellum* as including:

- Restoration (including return of sovereignty and re-entry into the wider community);
- Safeguarding the Innocent (particularly the impact on children, and also the effect of mines and chemical weapons);
- Respect for the Environment;
- Justice and Reconciliation; and
- Warrior Transition.

Wargaming Post Bellum

Given the nature of the *post bellum* phase, rigid wargames are probably not ideally suited for its exploration. However, as considered above, the USECT urban doctrine includes a consideration of the *post bellum* in its Consolidation and Transition phases, and the author's *City & CEMA* wargame includes these phases – although as discussions rather than map moves. There are a few wargames which focus on humanitarian operations – such as *Aftershock* (Brynen & Fisher, 2015) – which cover some of the issues faced *post bellum*, but otherwise the *post bellum* phase appears to be poorly served by wargames.

Within a less rigid matrix game the questions posed by Bass could serve to frame a more strategic consideration of the *post bellum* activities, whilst Iasiello's factors bring the discussion to a more operational level. Even more than with the *jus ad bellum* and *jus in bello*, there are few metrics that can be applied, and justice will be very much in the eye of each participant. As with all matrix games it is important that the participants role-play as much as possible and try and think themselves into the roles of the civilians, officials and others to fully explore the *post bellum* environment. Some research has even found that good role-playing can be just as, and sometimes more, accurate and informative than expert judgment. (Armstrong, 2001; Green, 2005).

Ideally wargames should be done before the hostilities are even started, in parallel with the *ad bellum* activities, to ensure that the war being contemplated does have a satisfactory conclusion to aim for, and that the planned action does not win the war but lose the peace. It is notable that much of the renewed interest in *jus post bellum* came after the US and Allied experience in Iraq (Iasiello, 2004). As with *jus ad bellum*, in *jus post bellum* many elements are likely to be more acute in urban environments due to the greater density of people and infrastructure. The cities are also likely to be the focal point for any discussions (or protests) regarding sovereignty, reconciliation, justice and other issues. Grey-Zone operations are unlikely to cease in the *post bellum* phase, with all sides are engaging in them to 'win the peace', particularly through the information war, and to set themselves up for the global realities and any new order that emerges beyond the conflict. As such perhaps one of the first steps, if a somewhat cynical one, on entering this phase should be a Grey-Zone wargame in order to understand how the actors are likely to try and shape and exploit the *post bellum* period.

AI AND *POST BELLUM*

The role of AI *post bellum* is likely to be quite different from other phases. AI ought to be being used in an engineering and social capacity, helping to manage and execute recovery and rebuilding plans, and to support the treatment of the harmed. Where there is a risk of insurgency after the war then there may be a more military use for AI. Biometrics were used to help manage post-war populations and identify insurgents amongst the peaceful populations in Iraq (Pendall & Sieg, 2014) and its use has also been suggested in Gaza (Spencer, n.d.). AI could similarly assist in the management of so-called gated communities, such as used in Iraq to segregate different sections of the population (Niva, 2008; Hristova, 2018).

AI, WARGAMING AND *POST BELLUM*

Wargaming, particularly through matrix games, could possibly provide a good means by which to surface some of the issues that the use of AI in such *post bellum* roles may face – if the players are representative of the community that will be on the receiving end of them. AI's contribution to *post bellum* wargaming is probably like that for *in bello*, being able to create large populations in agent-based models and modelling the responses, sentiments and attitudes of the population at large, and, with improving AI ability, being able to provide in-game dialogue with key civilian players.

WARGAMING ETHICS AND WARGAMING AI ETHICS

Closely associated with the use of wargaming within a just war context is that of the ethics of the wargaming process itself, and in the use of AI to support wargaming. Many of the issues of wargaming when related to the civilian experience of war may touch on dark topics, as explored in *Dark Places: Professional Wargaming of Dark Themes* (Curry, 2024).

One concern is what has been termed the 'indecency of recency', a disinclination to design or play wargames of combats where those affected are still alive (Train, 2016). However it has been rightly argued that 'recent conflicts are not only an appropriate topic for games, but perhaps the most appropriate' (Train & Ruhnke, 2016) and that 'games can't be too soon they can only be too late ... games are tools for understanding moments and there is no moment as important as our present moment because that's the moment that saves lives' (Wehrle, 2024). Hirst (2022) highlights concerns about the institutional approach to wargames and the need to ensure that they are conducted and used in an ethical manner and that any impact on those who use them is understood. Booth (2021), in the context of discussing informed consent, even talks about an '*Ender's Game* conundrum' – derived from the book and film of the same name, where the game players are not told that their game actions are real actions, resulting in real (and, in the story, genocidal level) deaths. The Derby House Principles are one grass-roots attempt to improve the ethics of wargaming from a diversity and inclusion perspective (PAXsims, 2020). A 2023 survey by Barzashka (2023) found that of 140 wargame designers '80 percent of the analytical wargames skipped ethics reviews' (Influence without oversight, para. 5). Key concerns of wargames in general are around the lack of explainability and implicit or explicit bias in the wargame design – manual wargames can help to avoid the former but not the latter, and Barazashka warns that the ethical challenges of wargames 'multiply as wargaming embraces AI' (Barzashka, 2023, The next frontier: Fusing AI and wargaming, para. 1). If wargames are effective at influencing decision-making then that influence needs to be on a sound footing.

CONCLUSION

Wargaming has become an increasingly valuable tool for militaries when developing force capabilities, training for war, and planning operations, in both its manual and computer-based forms. However, many wargame designs omit the elements which would make them of direct use in considering just war issues – particularly innocent civilians and reverberating damage.

The requirements of considering *jus ad bellum* are probably better addressed by less-rigid wargames, such as matrix games, where a wide range of issues and actions can be considered, although the fact that such games are typically adjudicated by the players, or a small team of experts, means that they are in danger of perpetuating biases. The issues of *jus in bello* are more amenable to examination with rigid games. However, most of these games focus too much on the actual fight, and not enough on the understanding and shaping actions or on the civilian harm and reverberating damage which are needed to help evaluate whether *jus in bello* is being achieved. As with *jus ad bellum*, *jus post bellum* is probably best served by less rigid games – or as a discussion phase after a rigid wargame since the considerations are less amenable to a numerical treatment. As US National Security Advisor Jake Sullivan said when talking about Gaza, 'there was no mathematical formula to decide if a plan was acceptable' (Wintour, 2024, para. 15).

AI is changing the character, if not the nature, or warfare, and so wargames need to be able to reflect these changes. This is possibly easier in abstracted manual games than it is in computer-based games, although computer-based games may have the option of incorporating real AI systems. AI can help across almost all the typical tasks involved in developing, executing and analysing a wargame. Its use to provide a better sense of the population, whether through synthetic social media environments or virtual participants in matrix style games, is possibly one of the biggest contributions it could make to better understanding just war issues through wargaming.

This chapter has shown how different types of wargame can contribute to an analysis and even evaluation of just war considerations. It has also highlighted that some games are already embedding game mechanics in their design to enable the exploration of just war issues. Moving forward it would be good to see more use of wargaming in considering just war issues in relation to current and potential conflicts, across all three dimensions of the just war. This could be further encouraged if more wargame designers included aspects related to the civilian population, collateral and reverberating damage, proportionality and necessity and even *mala in se* within their designs. Finally, the ethical issues around wargaming itself need to be considered, accepted or addressed. A just wargame is needed to support a just war.

My thanks to Evan d'Alessandro for comments and suggestions on an early draft of this chapter.

NOTE

1 Full Disclosure: Daden Limited is owned by the author and the author was the lead for this project.

REFERENCES

Allen, J., & Baker, D.-P. (2022). Can the Robots Save the City?: Ethics, Urban Warfare, and Autonomous Weapons. In *The Ethics of Urban Warfare* (pp. 172–186). Brill Nijhoff. https://doi.org/10.1163/9789004522404_011

Armstrong, J. S. (2001). Role Playing: A Method to Forecast Decisions. In J. S. Armstrong (Ed.), *Principles of Forecasting* (Vol. 30, pp. 15–30). Springer US. https://doi.org/10.1007/978-0-306-47630-3_2

Azad, T. M., Haider, M. W., & Sadiq, M. (2023). Understanding Gray Zone Warfare from Multiple Perspectives. *World Affairs*, 186(1), 81–104. https://doi.org/10.1177/00438200221141101

Barachini, F., & Stary, C. (2022). *From Digital Twins to Digital Selves and Beyond: Engineering and Social Models for a Trans-humanist World*. Springer Nature. https://library.oapen.org/handle/20.500.12657/54024

Barzashka, I. (2023, December 4). Wargames and AI: A dangerous mix that needs ethical oversight. *Bulletin of the Atomic Scientists*. https://thebulletin.org/2023/12/wargames-and-ai-a-dangerous-mix-that-needs-ethical-oversight/

Bass, G. J. (2004). Jus post bellum. *Philosophy & Public Affairs*, 32(4), 384–412.

Bellamy, A. J. (2008). The responsibilities of victory: Jus post bellum and the just war. *Review of International Studies*, 34(4), 601–625.

Bica, C. C. (1998). Interpreting Just War Theory's Jus in Bello Criterion of Discrimination. *Public Affairs Quarterly*, 12(2), 157–168.

Booth, R. (2021). *Playing on the Edge: Avoiding Ethical Pitfalls in Wargaming*. Sandia National Lab.(SNL-CA), Livermore, CA (United States). https://www.osti.gov/servlets/purl/1872518

Brands, H. (2016). *Paradoxes of the Gray Zone* (SSRN Scholarly Paper 2737593). https://doi.org/10.2139/ssrn.2737593

Brynen, R., & Fisher, T. (2015). *Aftershock*. PAXsims.

Buchanan, H. (2022). *Flashpoint: South China Sea*. GMT Games. https://boardgamegeek.com/boardgame/247182/flashpoint-south-china-sea

Burden, D. (2022, April 26). *Automated Generation of Information Injects*. IT2EC, London, England.

Burden, D. (2023a). The Battles of Hue: Understanding Urban Conflicts through Wargaming. *Journal of Strategic Security*, 16(3). https://doi.org/10.5038/1944-0472.16.3.2138
Burden, D. J. H. (2023b). *City & CEMA*. Tau Noyen Games.
Burden, D. J. H. (in press). A Typology for the Representation of Civilians within Wargames. *The Nugget.*
Caffrey, M. B., Jr. (2018, July 17). *Introduction to Wargaming*. Connections, National Defense University, USA. https://connections-wargaming.com/wp-content/uploads/2018/08/caffrey-introduction-to-wargaming-2018.pdf
Caffrey, M. B. (2019). *On Wargaming: How Wargames Have Shaped History and How They May Shape the Future* (Vol. 43). Naval War College Press.
Carment, D., & Belo, D. (2018). *War's Future: The Risks and Rewards of Grey Zone Conflict and Hybrid Warfare*. Canadian Global Affairs Institute.
Colonomos, A. (2015). Is there a Future for 'Jus ex Bello'? *Global Policy*, 6(4), 358–368. https://doi.org/10.1111/1758-5899.12244
Curry, J. (2024). *Dark Places: Professional Wargaming of Dark Themes*. History of Wargaming Project. http://www.wargaming.co/professional/details/darkplaces.htm
Curry, J., Engle, C., & Perla, P. (2011). *The Matrix Games Handbook*. History of Wargaming Project.
Department of Defense. (2013). *Joint Urban Operations, Joint Publication 3-06*. Department of Defense. https://www.jcs.mil/Portals/36/Documents/Doctrine/pubs/jp3_06.pdf
Dige, M. (2017). Explaining the Principle of Mala in Se. In *The Just War Tradition: Applying Old Ethics to New Problems* (pp. 141–155). Routledge. https://www.taylorfrancis.com/chapters/edit/10.4324/9781315086125-10/explaining-principle-mala-se-1-morten-dige
Geist, E., Frank, A. B., & Menthe, L. (2024). *Understanding the Limits of Artificial Intelligence for Warfighters: Volume 4, Wargames*. RAND Corporation. https://www.rand.org/pubs/research_reports/RRA1722-4.html
Green, K. C. (2005). Game theory, simulated interaction, and unaided judgement for forecasting decisions in conflicts: Further evidence. *International Journal of Forecasting*, 21(3), 463–472.
Gupta, A., & Matthews, J. (2005). *Twilight Struggle*. GMT Games. https://boardgamegeek.com/boardgame/12333/twilight-struggle
Hills, A. (2004). *Future War in Cities: Rethinking a Liberal Dilemma*. Routledge. https://www.routledge.com/Future-War-In-Cities-Rethinking-a-Liberal-Dilemma/Hills/p/book/9780714684949
Hirst, A. (2022). Wargames resurgent: The hyperrealities of military gaming from recruitment to rehabilitation. *International Studies Quarterly*, 66(3), sqac021.
Howard, N. (1999). *Confrontation analysis: How to win operations other than war*. Center for Advanced Concepts and Technology (ACT). https://apps.dtic.mil/sti/citations/ADA458371
Hristova, S. (2018). Charting the territory: Space and power in the Iraq War. *ACME: An International Journal for Critical Geographies*, 17(4), 939–957.
Hunter, C., & Bowen, B. E. (2023). We'll never have a model of an AI major-general: Artificial Intelligence, command decisions, and kitsch visions of war. *Journal of Strategic Studies*, 0(0), 1–31. https://doi.org/10.1080/01402390.2023.2241648
Iasiello, L. V. (2004). Jus post bellum: The moral responsibilities of victors in war. *Naval War College Review*, 57(3/4), 33–52.
Kerr, C. C., Stuart, R. M., Mistry, D., Abeysuriya, R. G., Rosenfeld, K., Hart, G. R., Núñez, R. C., Cohen, J. A., Selvaraj, P., & Hagedorn, B. (2021). Covasim: An agent-based model of COVID-19 dynamics and interventions. *PLOS Computational Biology*, 17(7), e1009149.
King, A. (2021). *Urban Warfare in the Twenty-First Century*. PolityPress.
Knack, A., & Powell, R. (2023). *Artificial Intelligence in Wargaming*. CETaS, Alan Turing Institute.

Le Ménahèze, J., Brewer, H., Train, B., & Brynen, R. (2017). *We Are Coming Nineveh*. Nuts! Publishing.

Longley-Brown, G. (2019). *Successful Professional Wargames: A Practitioner's Handbook* (Ed. J. Curry). The History of Wargaming Project.

Martinson, R. D. (2024). The PLA Navy's Blue Team Center Games for War. *Proc. US Naval Institute*, 150(5), 1455.

Mollendorf, D. (2008). Jus ex bello. *Journal of Political Philosophy*, 16(2), 123–136.

Mollendorf, D. (2015). Two Doctrines of Jus ex Bello. *Ethics*, 125(3), 653–673. https://doi.org/10.1086/679560

Mouat, T. (2022). The Use and Misuse of Wargames. *Scandinavian Journal of Military Studies*, 5(ue. 1), 209–220. https://doi.org/10.31374/sjms.121

Niva, S. (2008). Walling off Iraq: Israel's imprint on US counterinsurgency doctrine. *Middle East Policy*, 15(3), 67–79.

Ohlin, J. D. (2017). *Sharp Wars Are Brief.* Cornell Legal Studies Research Paper No. 17-13. https://papers.ssrn.com/sol3/papers.cfm?abstract_id=2931877

Oracz, M., & Wiśniewski, J. (2017). *This War of Mine*. 11bit studios. https://boardgamegeek.com/boardgame/188920/this-war-of-mine-the-board-game

Patterson, E., & LiVecche, M. (2023). *Military Necessity and Just War Statecraft: The Principle of National Security Stewardship*. Taylor & Francis.

PAXsims. (2020). *The Derby House Principles*. PAXsims. https://paxsims.wordpress.com/derby-house-principles/

Pendall, D., & Sieg, C. (2014). Biometric-Enabled Intelligence in Regional Command–East. *Joint Forces Quarterly*, 72(1), 70.

Perla, P. (2011). *Peter Perla's The Art of Wargaming* (Ed. J. Curry). The History of Wargaming Project.

Pettyjohn, S. L., & Wasser, B. (2019). *Competing in the Gray Zone (RR2791)*. Rand Corporation. https://www.rand.org/content/dam/rand/pubs/research_reports/RR2700/RR2791/RAND_RR2791.pdf

Schuck, M. J. (1994). When the shooting stops: Missing elements in just war theory. *Christian Century*, 111(30), 982–984.

Silva Simões, P. D., & Ferreira, C. G. I. (2011). Military War Games Edutainment. *IEEE 1st International Conference on Serious Games and Applications for Health (SeGAH)*, 16–18 November 2011.

Spencer, J. (n.d.). Defeating the Urban Enemy, with General David Petraeus. [Audio podcast episode]. In *Urban Warfare Project Podcast*. Modern Warfare Institute. https://mwi.westpoint.edu/urban-warfare-project-podcast-defeating-the-urban-enemy-with-general-david-petraeus/

Stahn, C. (2007). Jus post bellum: Mapping the discipline (s). *American University International Law Review*, 23, 311.

Stahn, C., Easterday, J. S., & Iverson, J. (2014). *Jus Post Bellum: Mapping the Normative Foundations*. Oxford University Press.

Tandon, V., & Mehra, R. (2023). *Study of Approaches to Predict Personality Using Digital Twin*. https://www.intechopen.com/online-first/86463

Train, B. (2016, March 22). Bored of War the modern irregular wargame as indecent, critical, subversive object. *National Conference of the Popular Culture Association, Seattle, USA*. https://brtrain.wordpress.com/wp-content/uploads/2016/03/bored20mar.pdf

Train, B., & Ruhnke, V. (2016). *Chess, Go, and Vietnam: Gaming Modern Insurgency*. (Eds. P. Harrigan & M.G. Kirscehnbaum) *Zones of Control*. https://direct.mit.edu/books/edited-volume/chapter-pdf/2263920/9780262334945_cbq.pdf

United Nations. (2004). *A More Secure World: Our Shared Responsibility, Report of the High-level Panel on Threats, Challenges and Change*. United Nations. https://www.un.org/peacebuilding/sites/www.un.org.peacebuilding/files/documents/hlp_more_secure_world.pdf

US Army. (1964). *FM 31-50 Combat in Urban and Built-Up Areas*. US Army.

US Department of Defense (2020). *Joint Planning, Joint Publication 5-0*. Department of Defense. https://irp.fas.org/doddir/dod/jp5_0.pdf

Vakas, D., Prince, J., Blacksten, H. R., & Burdick, C. (2001). Commander behavior and course of action selection in JWARS. *Proceeding of the 2001 Winter Simulation Conference (Cat. No. 01CH37304)*, 1, 697–705. https://ieeexplore.ieee.org/abstract/document/977357/

Van Steenberghe, R. (2012). Proportionality under Jus ad bellum and Jus in bello: Clarifying their Relationship. *Israel Law Review*, 45(1), 107–124.

Walzer, M. (2015). *Just and Unjust Wars: A Moral Argument with Historical Illustrations*. Hachette UK.

Wasser, B., Oberholtzer, J., Pettyjohn, S. L., & Mackenzie, W. (2019). *Gaming Gray Zone Tactics (RR2915)*. Rand Corporation. https://view.ckcest.cn/AllFiles/ZKBG/Pages/313/RAND_RR2915.pdf

Webber, P. (2021, February). *Modern Operational Level Naval Wargames: Design Challenges and Techniques*: Georgetown University Wargames Society, Washington, D.C. https://www.youtube.com/watch?v=5JbSn5bevnI

Wehrle, C. (2024, April 23). *Perspective and its Consequences in Historical Game Design*. Georgetown, USA: Georgetown University Wargaming Society. https://www.youtube.com/watch?v=wi_MpZxLPFM

Weuve, C. A., Perla, P. P., Markowitz, M. C., Rubel, R., Downes-Martin, S., Martin, M., & Vebber, P. V. (2004). *Wargame pathologies*. Newport: Naval War College. https://doi.org/10.21236/ADA596774

Wintour, P. (2024, May 24). US and UK to back Israel over ICJ ruling after blurring their Rafah red lines. *The Guardian*. https://www.theguardian.com/world/article/2024/may/24/us-and-uk-to-back-israel-against-icj-while-blurring-their-rafah-red-lines

10 Fighting a Just War in a Digital Realm

Zak Savin-Baden

The idea of war within the digital realm has always been a core section of the virtual worlds since its inception and the rise of the Call of Duty, the Halo, and the Grand Theft Auto franchises. Yet the concept of Just War has often been lost within these areas with the players not caring about the rules of morality or of war but solely focusing on improving their high scores. Furthermore, some of these games are limited in their game modes which means fighting a Just War is almost impossible. The argument of this chapter is that Just Wary Theory needs to be explored in depth within the gaming community. Further, throughout the chapter, the various ways in which Just Wars can be fought within the virtual spaces will be explored.

THE CONTEXT OF A JUST WAR IN THE DIGITAL REALM AND GAMING COMMUNITY

For the analysis of Just War in the digital realm, the existing conflicts and previous actions within the digital realm must be contextualised. This is done in the form of a literature review, that examines a variety of papers that discuss actions taken in the digital realm and their application to Just War Theory and is illustrated in Table 10.1.

The findings of the literature review focuses on the role of the Game Designers and Developers regarding Just War, arguing that they must ensure that the players have the option to conduct Just War. However, the approach generally taken is that players regardless of Game Features will find ways to fight both justly and unjustly such as the WoW players assaulting the virtual funeral as (Losh, 2009) describes. Furthermore, (Guanio-Uluru, 2016), discusses that wargaming, particularly in fiction such as Ender's Game and The Hunger Games are inherently unjust with diabolical tactics used. However, the conclusion is that peaceful games are generally more popular than wargames, implying that most gamers do strive for peace. However, the paper by (Dunn, 2012) discusses the ideas that the Virtual Realms must have different rules to the real world due to the Virtual Worlds sometimes sharing more similarities with dreams yet unlike dreams, the actions taken are significantly more real. Therefore, these worlds still require rules, yet care must be taken as to how these are applied. Regarding the paper by (Young, 2018), it covers a wide range of political actions covered in a variety of virtual worlds and discusses various cases of morality. However, the author concludes that due to certain limitations, it is more thorough to evaluate the media further when it is produced into more media formats such as Movies or TV.

 DOI: 10.1201/9781003440253-10

TABLE 10.1
Overview of the Literature

Article	Topic	Keywords	Title	Description	Value (/5)	Initial Analysis
Donald (2017)	Game Design and Just War	first-person shooter, just war, war crimes, video games, game design	Just War? War Games, War Crimes, and Game Design	A paper detailing the war crimes and civilian usage in games as well as the role of the player, particular regarding game design	4	The paper concludes that Game Designers and Developers must take a more active role in designing their games without war crimes or with the goal of condemning their use.
Guanio-Uluru (2016)	Just War, Fiction, War as a Game	war, games, ethics, narrative, Ender's Game, The Hunger Games, Prachett	War, Games, and the Ethics of Fiction	A paper discussing the ethics and the idea of Wars in Fiction such as Ender's Game and The Hunger Games	3	Though the paper lacks a focus on Just War, it does provide insight into the role that fiction and games take with it concluding that peaceful competitive games are sold more often than war games
Dunn (2012)	Virtual Worlds and Virtual Actions	Virtual, Virtual actions, Virtual worlds, Consent, Video games, Closed world, Play	Virtual worlds and moral evaluation	A paper discussing how morality should be applied within the virtual words	4	The paper concludes that the virtual worlds have different rules like dreams, yet the players consent regarding actions taken must help define the morality of the action
Young (2018)	Politics, Virtual Worlds	Popular geopolitics; popular culture; video games	Playing Politics – Warfare in Virtual Worlds	A PhD thesis detailing a vast amount of information regarding politics within virtual worlds	5	The paper is highly detailed discussing a variety of political actions, particularly how they can occur in Virtual Worlds
Losh (2009)	Just War, Virtual Worlds	Virtual Worlds, War Crimes, World of Warcraft, Ethics	Regulating Violence in Virtual Worlds: Theorizing Just War and Defining War Crimes in World of Warcraft	A book detailing various incidents within the game World of Warcraft	4	The book details various incidents such as the attack on a funeral of a player who died in the real world as well as the extermination of the gold farmers within the game to evaluate their morality and justification

JUST WAR IN PARTICULAR GAMES

Yet if the attention is turned to the more open-world Sandbox style games with a combat theme, then it becomes more thought- provoking as in some cases, the rules of the world are set by the players such as in Minecraft and Space Engineers, whilst others have more severe restrictions such as EVE: Online and Arma 3. Within these games (Minecraft and Space Engineers) the idea of a Just War becomes considerably more plausible and is often the goal of some the player bases within these communities

Space Engineers

When it comes to Space Engineers, there does exist a roughly universal ruleset of War which was developed by one of the more notable players within the community known as Glazton (Figure 10.1).

Offlining refers to attacking a player or group of players while they are offline and therefore cannot fight back while insiding refers to deploying an alternative account into the group chats and generally the in-game player faction in order to eavesdrop enemy communications, as well as steal ships and resources then transferring them to the faction that they hold their allegiance to. Furthermore, there are generally two additional unwritten rules and whilst not explicitly included in this. These rules forbid the following:

Molemanning – digging underneath base defences.
Hydromanning – flying repeatedly at a ship or base to disassemble it using the grinding tool.
While the ruleset is only considered guidelines, most of them are adhered to by many players.

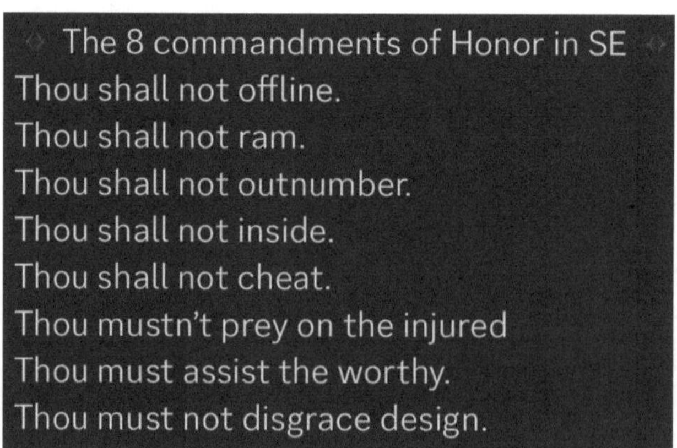

FIGURE 10.1 Just War Theory, the Space Engineers version.

The majority of rules are created and enforced by the administrators of player run servers. Therefore, a Just War must be fought with all these rules in mind as well as the server itself.

Furthermore, it must also be considered that within the Space Engineers, Civilians and the idea of Civilian bases and ships is difficult to define as most ships above a certain size are always armed to fend off Space Pirates whether NPC or Players. This is similar to Star Wars as even the Freighters such as the Millenium Falcon and Diplomatic Vessels such as the Tantive IV do possess weapons.

Yet some servers do account for this with Weapon Classifications, both to enforce fairer classes of warships and to allow suitable Civilian craft. The system of Weapon Classifications also brings onto the point of fighting a war fairly, as since all the warships are different due to being built by different people who have different definitions about armour and size of ships. One player's destroyer is another player's dreadnought. Furthermore, certain players have unique skills in designing and building ships that can outperform very similar ships by a significant margin. The point being made is that there are a lot of factors to consider and while on the surface, fighting a Just War in Space Engineers is easy from an outside perspective, it swiftly becomes more and more complicated to define when the nuances of the game are considered. There is an example below to demonstrate this.

The ARC faction was a small group of players that formed after the collapse of another faction called the IRS and had spent some time training and shipbuilding by another faction known as WCO who were notable for their huge, well-armoured and capable ships, being among the best. The ICP faction was a huge group of players with tried and tested designs and had been in existence for over two years by the point of the conflict (Table 10.2). Upon the ARC faction joining the server, they immediately began breaking and bending some of the server rules, one such example being offlining a base with no declaration and were foolishly caught on camera. They also harboured significant hatred towards the ICP who had been partly responsible for the collapse of the IRS, which resulted in being toxic in chat towards them. This later led to the odd skirmish or so until ARC eventually declared all-out war and stated their first target as Valiant, one of the ICP Space Stations. This first battle later became known as the Battle of Valiant Station.

The ICP Warships arranged themselves in a simple blockade formation around the base while the ARC warships formed a semi-circle and closed in. However, they met

TABLE 10.2
The Imperial Commonwealth of Planets – Alarian Rider Coalition Conflict

Faction Name	Abbreviation
The Imperial Commonwealth of Planets	ICP
Alarian Rider Coalition	ARC
Remnants of the Many	ROTM
World's Commonwealth	WCO
Inland Revenue Service	IRS

with a sheer wall of artillery fire, and they were forced to retreat after only one hour of fighting while typical battles generally average an hour and a half to around three and a half hours. Regarding the losses of this battle, ARC lost three ships with the rest suffering light to moderate damage while the ICP ships only took light damage.

At this stage is also important to consider the broader context. Firstly, each Factions' assets from an outside perspective may seem imbalanced. However, this was not the intention of the ICP who were generally known to fight fair. The logic behind bringing over double the firepower of ARC was that they had been bragging their brand-new destroyers were bleeding edge and due to the fact, their size was almost on par with the Drakon-Class Cruiser, this did have some degree of truth. Furthermore, the ARC ships were significantly more armoured than the ICP ships and were significantly more manoeuvrable. This is partly due to the design philosophy each faction followed with ICP building ships of various classes designed to work together in a fleet while ARC only built Frigates and Destroyers designed to operate on their own and fight a cruiser. Hence why the ICP brought a whole fleet to fight 6 'Cruiser-Killers' as shown in (Table 10.3). From this perspective, it's understandable as to why ICP brought so many ships and why this could be seen as fair. Unfortunately, due to both ARC overconfidence and ICP overestimating ARC's capability, it did result in the battle being slightly skewed in ICP's favour.

In conclusion with this battle, it can be argued it was fully just under the Just War conditions (Moseley, n.d.), as it was presumed that the forces were fair, the ICP were fighting with Just cause as they were defending themselves after failed diplomacy. Furthermore, the battle was supervised by the admins to ensure that it did not get out of hand and no civilian assets were destroyed or harmed. Finally, it led to a truce and the collapse of ARC's navy, paving the way for the Battle of Camp David where ARC were routed.

Yet regarding the Battle of Camp David, the war becomes more difficult to quantify as Just due to the numerous factors involved. Firstly, ARC's playerbase had declined leading them to recruit WCO. Secondly, ARC's navy was depleted, so they began constructing huge WCO warships. Thirdly, they had built enormous mining and welding rigs to assist in their building, which were banned as they caused significant server performance issues. Furthermore, ROTM, one of ARC's allies on the server who shared a friendship with both ICP and ARC, promptly defected to the

TABLE 10.3
The Battle of Valiant Station Breakdown

ICP Military Assets	ARC Military Assets
Valiant Space Station (Navy HQ)	6 Destroyers
1 Excalibur-Class Heavy Cruiser	
1 Drakon-Class Cruiser	
4 Fenrir-Class Fleet Destroyers	
2 Chrysaor-Class Fast Attack Destroyers	
3 Artemis-Class Frigates	

ICP. Regarding the battle, the ICP-ROTM coalition fleet annihilated the base. ARC panicked, failed to mount a defence then surrendered. The response was simply 'We are not currently accepting surrenders at this time' and continued their assault.

While the force was not proportional to the threat, one could argue that it would have been proportional had ARC had finished their fleet. Furthermore, although civilian assets were destroyed, said assets in question did violate server rules and were located at a military base at the time, therefore it could be argued that their destruction was just as no civilians were killed, furthermore the ICC provides further clarity in that as they were also being used for military purposes that they are fair game (ICRC, 2016). This battle ended the war and ARC left the server, ushering in a lasting peace for a few months (Table 10.4).

Overall, this example does demonstrate that a Just War can be fought to a certain extent, yet some issues of perspective must be considered. Especially since some Server Admins do allow the destruction of any asset which does break server rules providing it does not contravene the Offlining rules and provides valid evidence of the rule break.

The main consideration within this war however is the threat itself, as evil is difficult to define. ARC does fit a partial definition as they were toxic in chat and did break a significant number of rules as well as an offline attack. However, they did show restraint at times and were reasonable in some conversations when debating future server changes. Therefore, the main challenge is finding an enemy which is truly evil in which the war can indeed be somewhat genuine.

TABLE 10.4
Just War Evaluation of ICP and ARC Conflict

Just War Conditions	Battle Conditions
Just Cause	Self Defence against ARC then retaliatory Invasion to ensure ARC no longer invades or attacks anyone.
Declared and supervised by proper authorities	Declared by ARC and ICP and supervised by Server Admins.
Promote good and Justice as well as restore peace afterwards	A lasting peace was established after ARC left the server.
Last Resort when diplomacy fails	Yes
Proportional on both sides and civilian targets must be avoided	Proportional to begin with yet became skewed in ICP's favour after first battle while second battle, ARC only had half-complete military assets.
	Civilian Mining and Welding Equipment was destroyed but no civilians were killed and said Equipment in question did violate server rules and were stored at a military installation.
The good achieved must be greater than the evil before	ARC left the server, and a lasting peace was established.

Minecraft

Minecraft shares similar characteristics to Space Engineers, in the sense that there are general written and unwritten rules as well as the server rules which must be followed as well as there often being a significant lack of a 'Evil Empire' to fight. While some servers do indeed have full scale wars and battles against each other, most of these are scripted or at least semi-scripted such as in the case of the Dream SMP (Dream Team Wiki, n.d.). This does in a sense make Just Wars easy to play and to justify as it is reduced to a game or art to be consumed in the same way as a war movie. Yet there is a major outlier to this, the server known simply as 2b2t. This server known as Minecraft's oldest anarchy server, possessing no rules whatsoever and simply allowing the player base to play as they please which leads to a very high level of mistrust being players. The level of conflict and strife that is present on this server is on a truly colossal scale and there are various conflicts that could be covered, yet the one selected is the most famous, that being the Rusher War.

The Rusher War began when a Youtuber known as TheCampingRusher began playing on the server leading to a massive influx of his fans joining the server. A few days later on June 4th, 2016, TheCampingRusher declared war on all OG players, as in all the original players before him. Almost immediately some of the OGs mobilised, with a player known as FitMC leading a three-person team against the TheCampingRusher's fans, more commonly known as Rushers and later Team Rusher. FitMC and their team massacred all 40 players and destroyed the base. This led to FitMC becoming the leader of Team Veteran, a group of OGs who all allied to fight the Rushers, supported by some of the other factions on the server. The main strategy used by Team Veteran at the start was containment, aiming to contain all the Rushers within the main spawn area. Three different attempts were made, lava walls which failed when the Rushers built through or over them (The Lava Cast), flooding the spawn with water (Operation Poseidon) which failed due to the Rushers and surrounding environment managing to free the water, and perhaps the most ingenious, a massive bombing campaign using airships (Bombing of Spawn). Later in the war, one final attempt was made which saw the extensive use of suicide bombers to kill players and ended in disaster for Team Veteran due to the Rushers mass-adopting the strategy. This war caused considerable server issues causing the admin Hausemaster to temporarily institute a queue system and change to a different server map till problems could be resolved.

There were extensive campaigns beyond the previously stated conflicts, most involving Veteran and other smaller factions against the Rushers as well as a lot of backstabbing, infighting and political discourse which do make events murky. FitMC did engage with some personal duels against TheCampingRusher which FitMC generally won. The final battle came on September 16th, 2016, with a massive invasion of the spawn, initially just 40 Veterans v 3 Rushers but when others heard of the conflict, they came to join. The battle devolved into a massive free-for-all allowing TheCampingRusher to escape. After this TheCampingRusher left the server for two months and on the 9th of October 2016, FitMC declared the war over.

Overall, this does demonstrate how a war can be derailed and get out of hand very quickly when new concepts are developed quickly on the battlefield such as the use of the suicide bombers via End Crystals and Beds (end Crystals explode on contact whilst Beds explode when in a dimension), other than the Overworld to prevent players setting their spawn in other dimensions which plagued the server for long after the war ended. It also provides interesting insight into how a smaller power would attack a larger less-aggressive power. The opening tactics of the war are particularly key here as they are the most justifiable as it was simply based off containing the other side in a certain area where they could not mount any attacks successfully. Though a rather ambitious campaign, it was a valiant attempt to end the war quickly. Furthermore, this example of a war is a good example of how virtual wars can operate in the extreme level with no regulation whatsoever, with the only governance being the server owner and the morals of the players (Table 10.5).

In relation to Just War Theory, this highlights the need for centralised authorities to supervise the war to prevent the committing of atrocities as well as the protecting of civilians, particularly since a great many were killed who simply logged into the server to watch the fireworks. Furthermore, considering the constant suicide bombing and atrocities continued for months after the war had concluded, there shows a desperate need of both a centralised system and the importance of diplomacy when applied to Virtual Wars.

TABLE 10.5
Just War Evaluation of the Rusher War

Just War Conditions	War Conditions
Just Cause	OGs were defending themselves Rushers were just having fun
Declared and supervised by proper authorities	At later points of the War, FitMC, Hausemaster and TheCampingRusher did begin taking action when it began to get out of hand. E.g. Queue system
Promote good and Justice as well as restore peace afterwards	No, as the server continued with anarchy and an increased suicide bomber problem
Last Resort when diplomacy fails	No
Proportional on both sides and civilian targets must be avoided	It can be said to be somewhat proportional as Team Veteran had a mass of resources and experience to draw on while Team Rusher had pure numbers. Civilians were targeted as any players longing in for curiosity would have been presumed to be Rushers
The good achieved must be greater than the evil before	No due to the increased amount of suicide bombers invading spawn

Arma 3

With regard to Arma 3, it provides a much more realistic idea of how Just Wars would play out, the game itself being a highly realistic military simulator (Milsim) and it is often used to create and distribute fake news about ongoing conflicts such as the Russo-Ukraine War (Křižka and Bohemia Interactive, 2023). For the purposes of this section, the Antistasi Mod will be explored as it allows the players more freedom outside of the main campaign as well as putting the players in a more interesting setting, allowing them to play as a resistance group to take over the Island of Altis, starting off with minimal resources and having to fight a guerilla war against the government (BLUFOR) and an Invasion force (OPFOR) attacking Altis. The game begins slowly going from attacking patrol teams and destroying checkpoints to assaulting airbases and controlling towns. The players win by taking every air base to gain air superiority and by having more than 50% support from the Island Population. The Lose condition is simply if 1/3 of the total island population is killed as to prevent players simply lowering the island population to secure a quick victory (Moni et al., 2023).

This Mod on the surface is very interesting to evaluate according to the Just War Theory as it primarily starts with a Just Cause as the militia the player starts as is formed out of the occupants of Altis and is dedicated to the destruction of an Oppressive Government System and/or the Hostile Invaders attempting to invade. However, due to both powers being extensively corrupt, they cannot be classed as suitable authorities to supervise the war. Furthermore, both of these powers conduct failed diplomacy before the game starts yet as a militia, diplomacy is not offered as an option to the player so although diplomacy is technically conducted, it has negligible effect, therefore the Just War Criteria should be reclassified regarding Diplomacy. Further modification to the criteria should also be discussed when it comes to the forces of each side, as the player does start with a militia, requiring them to capture vehicles and bases off the opponents. However, one can argue that this is balanced as the other sides are ran by AI while the militia is run by the players. This becomes even more interesting as the AI frequently commits War Crimes and attacks Civilian Targets. Yet for the players this is a choice, for example you can commit war crimes by using Human Shields and Napalm as well as hitting Civilian Targets, but this does affect reputation scores within the game itself (Moni et al., 2023). Interestingly, the game has such an uphill curve at the start due to the level of forces rallied against you, it can be quite challenging to play the good guy yet as this goes on this get easier but not without issues. Due to Arma 3's level of realism, your character is a glass cannon, so it becomes more and more favourable to shoot first and ask questions later which can lead to some accidental civilian deaths when in deep foliage. Further problems can arise due disorientation in combat and the various weather conditions within the game obstructing communications which can lead to friendly fire. Regardless, if the players win, the Invaders retreat, and the player's militia takes over the government. Therefore, it can be said that there is a greater good upon winning yet this is difficult to evaluate due to the game ending at that point.

Overall, it is absolutely possible to fight a Just War within this Mod because so much more is in control of the players especially in regard to tactics and deployment of force and this relates to work by allowing more focused and specific approaches.

This is further enhanced by the focus on the people of the island who you must have on your side in order to win, therefore providing extensive encouragement to prioritise them.

However, it is important to remember this is the case with only the side ran by the players, yet this does give an interesting insight as to how an AI might respond to warfare. Although the AI within Arma is basic and simply doing what it's programmed to do in certain cases, whereas in real life it is likely to guess that AI advising tactics would draw upon data from previous wars. With the previous few wars generally having overwhelming force on one side, it is likely to be skewed towards bringing overwhelming force as well as using abominable tactics because Humans have used them previously such as (Stilwell, 2022)'s discussion on the US Military's usage on Agent Orange in the Vietnam War for clearing forest. Therefore future AI systems could draw on this tactic possibly without knowing the context and/or the health consequences.

EVE: Online

EVE represents a significant outlier from the games previously covered as the wars in EVE are mainly based on economics. Eve itself is based on creating a corporation and growing it to be the richest in can be. However, it's approach to war is fascinating, firstly as unlike the other examples, a fee must be paid every week of the war's duration to CONCORD (with some exceptions), a Non-Player Faction which essentially acts a rule enforcement system by summoning warships on the player or players committing felonies (wiki.eveuniversity.org, n.d.). In this regard, every war in EVE is regulated by a governing authority. This does have an exception with economic wars as it is theoretically possible to manipulate the markets and economies in EVE to push a corporation to near bankruptcy. However, this is a rather difficulty feat due to EVE's systems preventing you from going into debt. Furthermore, another unique feature that EVE implements whenever large-scale battles affect server performance is time dilation or bullet time which slows down 1 second of game time to 10 seconds of real time. This provides significant balance to both sides as it means that a greater balance is provided between the players with poor internet connections as well as reducing the lag which can favour either side (wiki.eveuniversity.org, n.d.). Despite these balances, EVE does allow significant freedom in strategy and troop deployment, therefore troops can be outnumbered, and a variety of tactics can be used. One such example of this is the Bloodbath of B-R5RB (EVE Online, 2014). This battle erupted when the Pandemic Legion and their ally N3 failed to make a territory payment, possibly caused by a game bug as they had both enough money and autopay system on. Their enemies, the CFC and the Russian Coalition decided to exploit this by taking the station and attempting to claim the territory. Both sides rallied their forces and pushed into the system. The battle raged for hours. The CFC also prevented reinforcements by blocking adjacent systems and when the N3/Pandemic Legion joint fleets attempted to retreat, they engaged interdiction fields to prevent it. (An interdiction field prevents ships from using warp speed/jumping to hyperspace or whatever FTL travel, that world has). Regarding the application to Just War Theory, the battle did not start with a Just Cause and rather due to a server bug where little to no diplomacy was conducted with all players preferring to shoot rather than talk. Despite this, it was supervised by the CONCORD system and no civilian targets were destroyed or damaged. Furthermore, due to the later capitulation of the Russian Bloc after this battle, it let to negotiation where N3 regain their territory and leading CFC to rise to the most dominant power, later rebranding as The Imperium. Therefore, it can be argued that a greater good was achieved than before the bloodbath occurred.

CONCLUSION

Fighting a Just War in a digital realm is definitively possible providing certain facts are true. The first being the game allows enough freedom for you to do so. Secondly, that the players in mind are committed to the rules in play. The latter fact is true of real life also. A main factor within the war is that War Crimes are frequently committed for several reasons. Firstly, for the memes and the banter as there is no repercussions and there is the freedom to so. Furthermore, there is almost an element of encouragement due to some games having features such as flamethrowers. Secondly, the use of War Crimes and overwhelming force is often used to bring a swift end to the war or to reduce the time the war takes. In any case, a long war is not preferable to any side despite being in the virtual realm as players often lose interest or must balance real life. So, while the digital realm does provide significant ease in waging a Just War, it is affected by the dedication of the players to the rules of war. Yet in most cases, the avoidance of civilian targets is generally not an issue as often they are not present on the battlefield, however this does impact the war itself as there is less need for restraint. Overall, due to the great nature of abstraction present within the digital realm and the removal of significant inhibitions, it is significantly easier to fight a Just War, however because of the gamified experience of the war, there is no real drive or motivation to do so, with most players either fixated on the win and/or having as much fun as possible.

REFERENCES

CONCORD - EVE University Wiki. (n.d.). Wiki.eveuniversity.org. Retrieved January 30, 2024, from https://wiki.eveuniversity.org/CONCORD

Corporation diplomacy - EVE University Wiki. (n.d.). Wiki.eveuniversity.org. Retrieved January 30, 2024, from https://wiki.eveuniversity.org/Corporation_diplomacy#Declaring_war

Donald, I. (2017). Just War? War Games, War Crimes, and Game Design. *Games and Culture*, 155541201772035. https://doi.org/10.1177/1555412017720359

Dunn, J. (2012). Virtual worlds and moral evaluation. *Ethics and Information Technology*, *14*(4), 255–265. https://doi.org/10.1007/s10676-012-9298-6

Guanio-Uluru, L. (2016). War, games, and the ethics of fiction. *Game Studies*, *16*(2). http://hdl.handle.net/10852/58768

ICRC. (2016). Frequently asked questions on the rules of war. Www.icrc.org; https://www.icrc.org/en/document/ihl-rules-of-war-FAQ-Geneva-Conventions

Křižka, P., & Bohemia Interactive. (2023, October 10). *Arma 3 footage being used as Fake News | Blog*. Bohemia Interactive. https://www.bohemia.net/blog/arma-3-footage-being-used-as-fake-news

Losh, E. (2009). Regulating Violence in Virtual Worlds: Theorizing Just War and Defining War Crimes in World of Warcraft. *Pacific Coast Philology*, *44*(2), 159–172. https://www.jstor.org/stable/25699564?saml_data=eyJzYW1sVG9rZW4iOiIyYzBhOGNkNC04ODgxLTQwZDAtOTBlZC0zMDFhOWQ1MTYwZTIiLCJpbnN0aXR1dGlvbklkcyI6WyJmMmYzYTJkNi0xYzFlLTRmMDYtYjNkYS03MGFkMDc1ZDU1NzQiXX0

Moni, Meerkat, Murphy, B., Giddi, DeathTouchWilly, & targetingsnake. (2023). *Airstrikes — Antistasi Guide documentation*. Official-Antistasi-Community.github.io. https://official-antistasi-community.github.io/A3-Antistasi-Docs/reference_guide/concepts_info/concepts/airstrikes.html

Moni, Meerkat, Murphy, B., Giddi, DeathTouchWilly, & targetingsnake. (2023). *Win / Loss Conditions — Antistasi Guide documentation*. Official-Antistasi-Community.github.io. https://official-antistasi-community.github.io/A3-Antistasi-Docs/reference_guide/concepts_info/concepts/win_loss_conditions.html#concept-winlossconditions

Moseley, A. (n.d.). *Just War Theory | Internet Encyclopedia of Philosophy*. Internet Encyclopedia of Philosophy. https://iep.utm.edu/justwar/#H2

Stilwell, B. (2022, August 9). *Why the US Used Agent Orange in Vietnam and What Makes It So Deadly*. Military.com. https://www.military.com/history/why-us-used-agent-orange-vietnam-and-what-makes-it-so-deadly.html

The Bloodbath of B-R5RB, Gaming's Most Destructive Battle Ever. (2014, February 1). EVE Online. https://www.eveonline.com/news/view/the-bloodbath-of-b-r5rb

Time dilation - EVE University Wiki. (n.d.). Wiki.eveuniversity.org. Retrieved January 30, 2024, from https://wiki.eveuniversity.org/Time_dilation

Various. (2023a, February 1). *Queue*. 2b2t Wiki. https://2b2t.miraheze.org/wiki/Queue

Various. (2023b, December 2). *Rusher War*. 2b2t Wiki. https://2b2t.miraheze.org/wiki/Rusher_War

Various. (n.d.). *Timeline - 2b2t Wiki*. 2b2t.miraheze.org. https://2b2t.miraheze.org/wiki/Timeline

Wartime Operations in EVE University - EVE University Wiki. (n.d.). Wiki.eveuniversity.org. Retrieved January 30, 2024, from https://wiki.eveuniversity.org/Wartime_Operations_in_EVE_University

Wilbur Soot. (n.d.). Dream Team Wiki. Retrieved January 30, 2024, from https://dreamteam.fandom.com/wiki/Wilbur_Soot

Young, R. J. (2018, May 1). *Playing politics – warfare in virtual worlds*. https://etheses.whiterose.ac.uk/22035

Index

Pages in *italics* refer to figures and pages in **bold** refer to tables.

N

O

P

Q

R

S

T

U

V

W

Y

Z

FSC
www.fsc.org
MIX
Papier | Fördert gute Waldnutzung
FSC® C083411

Kolibri 360 GmbH
Ferdinand-Jühlke-Straße 7
99095 Erfurt, Deutschland
produktsicherheit@kolibri360.de